IF YOU SUFFER FROM EVEN *ONE* OF THESE SYMPTOMS, YOU MUST READ THIS BOOK. *BECAUSE NOW THERE'S HOPE...*

"I'm tired when I go to bed at night and just as tired — if not more — in the morning. Sleep doesn't seem to do much good."

"When I'm under stres of my neck, or the fron Most often, the muscl tense, tender, and painf

"My joints, especially swollen."

"My hands and feet tingle a lot, as if I'm getting electric shocks."

"I don't handle stress very well. Sometimes I can see a direct relationship between my muscle aches and pains and this kind of tension, but a lot of times the aches come when I seem pretty relaxed."

THESE ARE THE WORDS OF PEOPLE WHO DISCOVERED THAT THEY HAD FIBROSITIS AFTER HAVING BEEN TOLD THAT THEY HAD A LESS TREATABLE, LESS CURABLE, AND LESS *PREVENTABLE* DISEASE. THEIR LIVES HAVE ALL CHANGED FOR THE BETTER. AND SO CAN YOURS. FIND OUT WHY A DIAGNOSIS OF CHRONIC MUSCLE PAIN SYNDROME IS THE BEST NEWS YOU COULD POSSIBLY HAVE — BECAUSE *YOU* CAN CURE YOURSELF — QUICKLY AND SIMPLY!

CHRONIC MUSCLE PAIN SYNDROME

HOW TO RECOGNIZE AND TREAT IT —AND FEEL BETTER ALL OVER!

Paul Davidson, M.D.

BERKLEY BOOKS, NEW YORK

Grateful acknowledgment is made to the following for permission to reprint previously published material:

The New England Journal of Medicine: excerpts from "Paradox of Health" by
Arthur Barsky from *The New England Journal of Medicine*, Vol. 319, p. 378.
Copyright © 1988 by Massachusetts Medical Society.

The New York Times: excerpts from "Fatigue 'Virus' Has Experts More Baffled
and Skeptical Than Ever" by Philip M. Boffey from the July 28, 1987, issue of
The New York Times. Copyright © 1987 by The New York Times Company.
Reprinted by permission.

Illustrations by Rebecca Etelamaki.

This Berkley book contains the complete
text of the original hardcover edition.

CHRONIC MUSCLE PAIN SYNDROME

A Berkley Book / published by arrangement with
Villard, a division of Random House, Inc.

PRINTING HISTORY
Villard edition / February 1990
Published simultaneously in Canada by
Random House of Canada, Limited, Toronto
Berkley edition / June 1991

ISBN: 0-425-12775-3

A BERKLEY BOOK ® TM 757,375
Berkley Books are published by The Berkley Publishing Group,
200 Madison Avenue, New York, New York 10016.
The name "BERKLEY" and the "B" logo
are trademarks belonging to Berkley Publishing Corporation.

PRINTED IN THE UNITED STATES OF AMERICA

10 9

And wearisome nights are appointed to me.
When I lie down, I say, When shall I arise,
and the night be gone? and I am full of tossings
to and fro unto the dawning of the day.
Job 7:3–4

Days of affliction have taken hold upon me.
My bones are pierced in me in the night season:
and my sinews take no rest.
Job 30:16–17

CHRONIC
MUSCLE PAIN
SYNDROME

ACKNOWLEDGMENTS

This book is in great part a synthesis of observations and reasonings that were recorded by hundreds of dedicated and caring physicians over the course of many centuries. Without these written legacies, the preparation of this book would have been impossible. Added to their offerings were the contributions and writings of many others who were concerned about understanding and relieving human pain and suffering. Among them number people from such diverse fields as psychology, physiology, neurophysiology, physical therapy, and massage therapy, as well as the Eastern and Western philosophers who taught us so much about the unity of mind and body. A complete list of those to whom I am indebted would be almost interminable.

In addition, I would like to extend my gratitude to

my patients, from whom I have learned so much. It was they who eventually helped me separate the wheat from the chaff of the many beliefs and theories regarding the aches and pains of fibrositis.

My thanks to Suzanne Lipsett for both her editorial suggestions and her insight. Also, my appreciation to Gryta Coates for her assistance in adding the true Yoga influence to the stretching exercises.

And to my wife, Ursula, and children, Daniel and Julie, my special thanks for their limitless patience and support.

CONTENTS

INTRODUCTION

If you suffer from chronic muscle aches, pains, and
stiffness, you may have a condition known as fibrositis.

Fibrositis is very common and yet quite unique.
Within the past few years this condition has become
recognized by rheumatologists as the second or third
most common rheumatic complaint seen by physicians.
It affects literally millions of people in all walks of life.
Fibrositis is unique among rheumatic complaints in
that to date no inflammation has been found to account
for it. Its causes may range from simple muscle imbalance or overuse to an interaction between body and
mind, or, to use the technical terminology, between
soma and psyche. In addition, fibrositis is unique in
that its successful treatment eventually depends on the
patient's understanding of the disorder as much as or
more so than the physician's.

Despite the astronomical numbers of fibrositis sufferers, relatively little of what is known in the medical community about the disorder has become common knowledge. I suspect that a partial explanation has been benign neglect on the part of the medical profession, a neglect fostered by the mistaken belief, now eroding, that fibrositis either did not exist or was simply to be treated with a few aspirin and a pat on the back. Today we know otherwise: Fibrositis does exist, and its aches and pains are very real.

Depending on its causes, the treatment of fibrositis may involve anything from very simple to very complex measures. Fibrositis can have many causes, and its treatment is often complicated by missing bits and pieces of a patient's history that the physician must discover in order to be helpful. Nevertheless, armed with answers to the right questions, and with patience and persistence, we can treat fibrositis very successfully, markedly relieving or eradicating the muscle aches and pains that are its primary symptoms. But to achieve any degree of success, it is absolutely essential that both the patient and the physician have an understanding of the elements that combine to produce the disorder.

The purpose of this book is to provide that understanding by taking you on a fibrositic odyssey. This trip spans 2,400 years, touches down in several countries, and exposes you to many concepts. During the journey, you will learn that the symptoms of fibrositis are real, that they are shared by many others, that you need not rely on medications for the rest of your life, and that the success of the treatment often lies mainly in your own hands.

You will also learn the history of fibrositis—a history that has been steeped in controversy. This controversy has led to so much unnecessary confusion in the medical profession that until recently medical textbooks in this country didn't mention or even denied the existence of fibrositis. Formerly not even a part of the curriculum in most of our medical schools, fibrositis even now is generally a neglected subject. As a result of this confusion and neglect, physicians have known little about it and have displayed a hesitancy to accept its symptoms as being of much significance. All too often, patients with fibrositis have been left with increasing anxiety, frustration—and muscle pain.

Today's more enlightened medical community recognizes fibrositis to be not only very prevalent among the general population but also an enormous factor in disability and work absenteeism and thus the basis of an immense economic drain in industrialized societies. In Australia, for example, fibrositis and related conditions, such as repetitive strain injuries, have recently been termed the new industrial epidemic. Despite this heavy impact on our society, however, fibrositis remains virtually unknown to the lay public and is only now coming to the attention of those who are responsible for the medical care of industrial workers. The Australian experience, as well as that of other countries, suggests that this book will be of value to physicians and nonphysicians alike, who are becoming involved with the prevention and treatment of the rheumatic problems associated with our industries.

My intent is not only to explain fibrositis in plain terms but also to distinguish it from the other known rheumatic conditions, and to help you evaluate whether or not your everyday aches and pains may be fibrositis. I will tell you what is presently known about the causes and again why this knowledge is essential to you if your fibrositis is to be successfully treated.

Later chapters present today's most effective methods of helping you get rid of these muscular miseries. The treatment program I have devised, called RETRAIN, emphasizes a natural approach and incorporates minimal or no drugs and medicines, depending on specific circumstances. RETRAIN directs itself to the causes of fibrositis as well as its symptoms. It assists you in recognizing the diverse factors that play important roles in your aches and pains—adverse external and internal stress, muscle strain, muscle fatigue or imbalance, concern or fear, anxiety, or injury, to name a few. Above all, RETRAIN is a flexible, adaptable program. It emphasizes the treatment of you as a person— a combination of body and mind, not just a collection of muscles and other tissues and organs. The elements of this program have been used successfully to treat thousands of fibrositis sufferers.

This book is dedicated not only to sufferers but to the many physicians who care for patients with chronic muscular aches and pains. The knowledge that we physicians use to diagnose and treat fibrositis has been scattered about in the medical literature for almost 150 years, and much of it tends more to obscure than enlighten. I hope to make it possible now for physicians and fibrositis sufferers to work in a closer partnership for treatment.

Finally, this book is meant to help those who do not suffer from fibrositis but who know and care about people who do. The intent is to help them realize that these pains are real and not simply imagined. Too often a sense of helpless frustration causes family members and friends of fibrositis sufferers to react unintentionally in a manner that magnifies the problem. Thus they themselves become unintended participants in a vicious circle of pain, fear, and discomfort. Understanding can reduce their helpless frustration and enable them to play a positive role in treating this disorder.

CHRONIC MUSCLE PAIN SYNDROME

CHAPTER

1

Fibrositis and You

THE BASICS

WHAT THIS BOOK IS ABOUT

This is a book about muscle aches, pains, and stiffness —not the temporary feelings you would expect after exercising or those that accompany the common cold or flu, but rather aches, pains, and stiffness that may occur for no apparent reason. It's about muscles that ache when they shouldn't, muscles that become unexpectedly tired and stiff, muscles that remain painful long after work or exercise is over, and muscles that should have healed soon after minor or even major injury but continue to cause misery and discomfort. It's about muscles that can ache all over your body or just in a localized area such as an arm or shoulder. It's about a condition that can affect both sexes but for an as yet unexplained reason strikes women far more than men. In short, this book is about a malady we call fibrositis. Consider these case histories:

I saw Gerry, a thirty-seven-year-old woman, in my office five years ago. A year before she came to me, she began having severe muscle aches and pains primarily in her legs but also in her neck and shoulder areas. Before the onset of her symptoms, she had been a very healthy woman, active in dancing and the theater in addition to holding various jobs. She described the pains as "like being stepped on."—they were deep and flulike. Her sleep was often interrupted by the pains, and she said she could hardly walk in the morning because of pain and stiffness. She consulted many physicians, including an internist, an orthopedic surgeon, a neurologist, and a neurosurgeon. Multiple laboratory studies and X rays were completely normal. Many diagnoses were considered, including arthritis, an obscure muscle disease, hormonal imbalances, and —of most concern to Gerry—multiple sclerosis. The uncertainty of her future added to her distress and, as she put it, "I was in a vicious circle of fear and pain." She was treated with a variety of painkillers and anti-inflammatory drugs that gave her no relief at all. She felt that no one, including her physicians, husband, and friends, believed she had the pains. At one point, she thought it might be "something in the water" and had her tap water at home monitored for various toxins by the water department. At times she began to doubt her own sanity. Finally, when she thought she experienced very slight improvement from cortisone pills, she felt that maybe it was arthritis and that she should see an arthritis doctor. Gerry's proper diagnosis was fibrositis.

Jim, a thirty-four-year-old man, was a successful writer of stories of wildlife and the outdoors. He spent much of his time hiking in the woods and mountains. His problem began with an ankle strain that limited his walking. Within weeks after his injury, he began to experience widespread muscle pains in his upper back, hips, and legs. Not one to run to the doctors, he tried rest and aspirin, only to find his pains increasing. He finally consulted two different physicians who suspected that he had an early form of arthritis despite normal blood tests and X rays. But the usual antiarthritic medications were useless to him, and the fear of losing his livelihood resulted in insomnia and depression. For fourteen months he lived under a cloud of anxiety, fearful of permanent crippling. Jim did not have arthritis, nor did he have a disease that would cripple him and lead to a loss of his occupation. He had fibrositis.

AN OVERVIEW OF FIBROSITIS

Case Studies

Annette, a thirty-two-year-old married woman, tried to hold back her tears as she spoke to me. She was in pain, and she was frightened. She had been suffering from severe muscle stiffness, aches and pains, and increasing fatigue for eight years, beginning a year after her marriage. Annette lived with her husband and two children in a small California town with "little intellectual stimulation," so she drove long distances for sculpture lessons and a literature course. Because of "the poor school system" she gained per-

*mission to tutor her young daughter at home. She was
on two church committees and in the choir. She was,
according to her husband, "a real dynamo." She had
been to many physicians and had received diagnoses
ranging from "probable rheumatoid arthritis or
lupus" to "hypochondriasis." She had many tender
muscles and complained of increasing fatigue but
never developed any abnormal laboratory or X-ray
studies. Treatment with various medications had not
only given her no relief but had seemed to make her
symptoms worse. Her sleep was constantly interrupted by visions of increasing and permanent disability.*

*What was her problem? Her true diagnosis was fi-
brositis. By utilizing three features of the RETRAIN
program—education, appropriate time for relaxation,
and muscle stretching exercises—Annette gradually
improved to the point where she had only mild discomforts, which she herself termed insignificant.*

The next patient's story could have been a replay of
Annette's. However, it was resolved in a much shorter
time.

*June was twenty-one years old. She had begun work-
ing in an attorney's office at a computer a few months
before I first saw her. Within two weeks of beginning
her job, she experienced the onset of muscle stiffness
and pains in her neck and shoulders, and insomnia.
These shortly became so severe that she began to feel
she was becoming disabled and would have to quit
her job. Her physician suspected she had rheumatoid
arthritis and prescribed rest, an anti-inflammatory
drug, and physiotherapy. Her pains worsened and
soon depression was added to her list of symptoms. In*

talking with her, I learned that she had been an aerobics exercise instructor before she got married and that her new job was completely sedentary. In order to be a "good secretary," she was actually working more than eight hours a day, most of which was spent sitting in front of a desk and staring at a computer. Her history, physical examination, and studies indicated that the diagnosis was fibrositis. Treatment consisted simply of two elements of the RETRAIN program. She was given an explanation of fibrositis and a vigorous activity and muscle-stretching program. June's symptoms disappeared in two weeks.

Defining Fibrositis

Generally speaking, there is a direct relationship between our knowledge of a disease or disorder and our ability to successfully treat it. Fibrositis is no exception. If you have fibrositis, the better you understand it, the more likely it is that you will be able to rid yourself of your aches and pains or reduce them to a tolerable level. It will help your understanding if we start with a general overview of what this condition is —and what it isn't—about.

The term *fibrositis* refers to a group of muscular and fibrous-tissue symptoms that has surely plagued mankind for centuries and probably for millennia. This group of symptoms has been given a host of other names over the past 150 years—muscular rheumatism, fibromyalgia, fibromyositis, tension myalgia, and pain amplification syndrome, to name only a few. The major symptoms of this disorder are chronic muscle aches, pains, and stiffness, often accompanied by a sense of fatigue and a lack of feeling of well-being. Note that I said chronic muscle aches, pains, and stiffness. Most physicians would be hesitant to consider the diagnosis of fibrositis until the symptoms were present for at least a few months. Symptoms due to transient viral infections or excessive exercise, for example, can mimic fibrositis to some extent, but these symptoms pass within a relatively short period of time.

A Syndrome

Fibrositis, in medical terminology, is a *syndrome* rather than a disease. A syndrome is a collection of symptoms or characteristics that constitute a disorder. Another example is premenstrual syndrome, which affects some women prior to menstruation and consists of such symptoms as irritability, tension, headaches, depression, fatigue, breast swelling or tenderness, abdominal bloating, and weight gain. In the case of fibrositis, the major symptoms of the syndrome are muscle aches, pain, stiffness, fatigue, and muscle-tender points in specific areas.

A Subjective Malady

One of the hallmarks of fibrositis is its lack of objective findings—the absence of anything abnormal as revealed by the methods doctors routinely use to explain and understand disease. In pure fibrositis, laboratory and X-ray studies are typically normal. Biopsies of muscle and other tissue under a microscope again show normality. The only consistent finding on physical examination is the presence of tender points in many muscles and in areas where tendons insert onto bone. Since only the patient, and not the examiner, experiences this discomfort, the findings are said to be sub-jective—that is, there is no present method of seeing or quantifying the findings or verifying the tenderness in any even moderately accurate way. In short, the examiner has to take the word of the patient that "it hurts; it's tender."

For far too long—centuries, in fact—the unexplained pain and discomfort associated with fibrositis were erroneously ascribed to inflamed tissues or arthritis. Or they were lumped in the category of psychosomatic disease and ignored or minimized by the mainstream medical community. Owing to the lack of specific objective findings, these pains were dismissed as imaginary or hypochondriacal or as minor inconveniences or nuisances. Two major developments changed that line of thinking. The first was the gradual emergence of the field of rheumatology, which is devoted to the study of arthritis and rheumatism. By empirical, epidemiological, and statistical methods, the experiences of rheumatologists in many countries have been compared and fibrositis has emerged as something "real" and very widespread, and the ranks of the nonbelievers in the medical field have steadily shrunk. The second was the recognition by industry that these so-called imaginary, minor, or inconvenient discomforts were costing many millions of dollars in lost work hours and disability payments. Thus fibrositis showed itself to be not only among the commonest causes, if not the most important cause, of musculoskeletal pain in the general population, but also high on the list of the most costly industrial disability conditions. Nothing is guaranteed to gain attention more quickly than something that drains money from industry and the world economy.

Gender

In general medical practice, the incidence of fibrositis appears highest by far in women. Most reports from physicians' offices and university clinics indicate that of fibrositis sufferers, 75 percent or more are women. The statistics from industry show the same pattern.

Age

To many people the very name *rheumatism* evokes a mental picture of an elderly person with aches and pains hunched over a cane. With regard to fibrositis, nothing could be further from the truth. In my own practice, most of the fibrositis sufferers I see are between twenty and fifty years of age. This age spread has been noted by most of my colleagues and has been documented many times over in the medical literature. Fibrositis, then, is not a condition found primarily in older people. It is most common in those who are young or middle-aged and in the most productive years of their lives.

Fibrositis has just recently been found to occur in children. A report in 1985 by two rheumatologists documented the diagnosis of fibrositis in a group of thirty-two patients from nine to seventeen years of age.[1] Interestingly, only one of these patients had been correctly diagnosed before being referred. More than half had no specific diagnosis at all, and thirteen of these patients had been previously diagnosed incorrectly as having an arthritic condition. Many of these youngsters, like their adult counterparts, had undergone very extensive and very expensive medical studies that could have been avoided had the true diagnosis been suspected.

The Causes

What do we know about the cause of fibrositis? A partial answer is that speaking of the cause will probably get us nowhere. Fibrositis appears to result from a confluence of causes. And how do we know that? The answer lies in the experiences of many hundreds of physicians who have observed and documented thousands of fibrositis cases. The evidence tells us that fibrositis results from a combination of factors that start with a definite susceptibility to it on the part of the sufferer. Other elements are environmental and psychological components that interact in a specific manner. The causes are, in short, numerous and involve environment plus body and mind, or soma and psyche.

Treatment and Prevention

Can fibrositis be successfully treated? The answer is an unqualified yes. Enough is known about fibrositis at this time to permit highly effective treatment in most cases. Furthermore, enough is known about fibrositis to permit effective *self*-treatment to some extent, or at the least to involve the patient in the course of action to a degree unusual in conventional medical practice. Self-treatment has no small significance in this age of enormous medical costs. Whatever sufferers can do to participate in their own treatment represents a great source of self-satisfaction and a saving of consumer dollars.

Not only can fibrositis be adequately treated, it can

be *prevented*. The same approaches to treatment, if applied early enough, can prevent its occurrence. This book will provide you with the knowledge you need to avoid these painful miseries, to give your muscles increased relaxation, and to increase your sense of well-being.

What Fibrositis Is Not

Now for a few words on what fibrositis is not. We'll be going into some more detail about this later in the book. Suffice it to say for now that fibrositis is not arthritis. Arthritis affects joints and can be due to inflammation, degeneration, infection, and a host of other afflictions. Fibrositis does not—and I repeat—does not cause any damage to the joints at all. It does not cause the joints to swell, and it does not deform them. Of course, arthritis is common, as is fibrositis, so it is not at all surprising that the two often accompany each other. In fact, arthritis in certain areas may predispose to fibrositis—particularly in the neck and lower back regions. We should, however, never lose sight of the fact that fibrositis and arthritis are two different problems and must be treated accordingly as such.

BEYOND MUSCLE PAINS: OTHER SYMPTOMS

The picture of fibrositis is certainly dominated by chronic muscle aches and pains, but those are often not the only symptoms that can cause misery and discomfort. If you have fibrositis, you may also suffer from one or more of the following problems:

- fatigue and exhaustion
- irritable bowels
- tension or migraine-type headaches
- sleep disturbances
- feelings of joint swelling
- feelings of numbness and tingling in your extremities
- tension and poor stress tolerance
- temperature- and humidity-sensitive symptoms
- anxiety and depression

Although these symptoms are not uncommon in the general population, they are found with much higher frequency in patients with fibrositis. Recognizing this association alerts us to some important clues, one of which is that if we look at fibrositis as a purely muscular problem, our view of this disorder is going to be far too narrow. In addition, these symptoms can help us point to and unearth some of the causes that contribute to the entire syndrome. Let's take a closer look at these other body signals that seem to be saying, "Something's not quite right here."

> NOTE: These symptoms often accompany fibrositis and may be associated with it in a benign way. However, they should not automatically be considered to have no other relationship or cause, and you should consult your doctor if any of them persist.

Fatigue and Exhaustion

"I don't understand it. I'm tired when I go to bed at night but just as tired—if not more so—in the morn-

ing when I get up. Sleep just doesn't seem to do me much good."

"I'm totally exhausted—no energy or pep. I don't even want to meet anyone new anymore."

"I just don't feel well. I'm listless and have no interest in sex, which is causing quite a problem between me and my husband. I think I've had every medical test in the book, and would you believe it—they're all normal."

The fatigue and exhaustion that may accompany fibrositis can be all-pervasive. It may be associated with disinterest in meeting people, a decrease in sexual desires, and a general lack of a feeling of well-being. More than 90 percent of fibrositis sufferers have been noted to suffer from daytime fatigue.[2]

Irritable Bowel Syndrome

"My bowels haven't been right for years. Either I have loose movements or I'm constipated. Also, I have a lot of gas, distension, and some cramps in my lower abdomen. I've seen two specialists, had a lot of tests, and have been told there's no inflammation or cancer."

The *irritable bowel syndrome (IBS)* is a group of symptoms resulting from hyperactivity of the bowels, primarily the colon. Except for abnormal motility of the colon—either too much or too little—no underlying disease process has been found to account for it. In one study it was found that 30 percent of the patients with fibrositis suffered from the irritable bowel syndrome.[3]

Tension and Migraine Headaches

"Whenever I'm under stress, I get this headache that can be in the back of my neck, or the front, or sometimes all over my head. Most often, the muscles in the back of my neck seem tense, tender, and painful. I've had them for a long time, and I've been told by a few doctors that they're tension headaches."

"My headaches can be on one side or the other, or sometimes on both. They're what you might call 'real sick headaches'—the light bothers my eyes, and I often get sick to my stomach and can't hold anything down. They've been diagnosed as migraine headaches."

Both tension and migraine headaches may accompany fibrositis. In a report and review in the *Journal of the American Medical Association* in 1987, Dr. D. L. Goldenberg found that headaches were a problem in as many as 55 percent of fibrositis sufferers.[4] In my experience, the tension headaches, associated with muscle tension and tenderness in the muscles of the posterior cervical, or back of neck, area, are the commoner of the two by far.

Sleep Disturbances

"I can get to sleep okay, but I awaken a lot during the night with pain and aching in my muscles. Sometimes I feel like I've been run over by a truck."

"It's really difficult for me to fall asleep. All sorts of things run through my head. I think of what I have to do tomorrow and what I didn't do today. The more I

think, the more I ache. When I get up in the morning, it's like I never slept at all."

Again, sleep disturbances are common accompanying problems of fibrositis. One study revealed that up to 75 percent of fibrositis patients had this complaint.[4] Other studies suggest that sleep abnormalities play an important role in the production of muscular discomforts and fibrositis.[5]

Feelings of Joint Swelling

"My joints, especially my finger joints, feel like they're swollen. Sometimes my rings are hard to get on or off. I've been to four doctors and they all tell me they can't see any swelling of the joints, and my X rays are always normal."

About 50 percent of fibrositis sufferers experience the sensation of swollen joints when indeed no swelling is actually present.[4] In the general population, complaints of hand swelling that is not limited to the joints are most common among women. Some fluid retention on a hormonal basis (estrogen effects, in particular) occurs particularly in warm weather, and its frequency suggests that it is probably a normal phenomenon. The marked feeling of joint swelling that is so frequently noted with fibrositis is therefore a different complaint. If indeed joint swelling is present (which it is not in fibrositis), some other process, such as arthritis, must be considered.

Feelings of Numbness and Tingling

"My hands and feet tingle a lot, as if I'm getting little electric shocks. Sometimes the tingling goes down my

arms or legs. It's really annoying. It can come and go, with no pattern that I can figure out. My hands and feet also seem numb, as if I can't feel anything, but I really can. It sometimes helps to get up and walk around."

Numbness means different things to different people. Many people interpret tingling or the feeling of one's extremities "falling asleep" as numbness. To the physician, however, true numbness means the absence of feeling in an area of skin; the skin actually feels dead when touched by cotton or pricked by a pin and transmits no tactile stimulation. An example is the numbness that occurs when the dentist injects your mouth, or the surgeon injects your skin, with a local anesthetic such as lidocaine. Numbness indicates that a nerve to a specific skin area is temporarily asleep or has suffered an injury. This is not the case in fibrositis. Again, the tingling and numbness that accompanies fibrositis is subjective—felt by the patient, with no detectable abnormality, even with sophisticated nerve-conduction studies. Anywhere from 50 to about 90 percent of fibrositis sufferers have this complaint.[3,4]

Tension and Poor Stress Tolerance

"I don't handle stress very well. As a kid, I used to get pretty uptight when I had to take tests at school. I did well, and I still do well, but it takes its toll on me. Now, for example, I tense up when I have to speak before a group, or when my boss gives me the least bit of criticism. It bothers me, because I want to do a good job, but I wonder if I'm really doing it. I've tried tranquilizers, but in the long run they seem to make

me worse. Sometimes I can see a direct relationship between my muscle aches and pains with this kind of tension, but a lot of times the aches come when I seem pretty relaxed."

For many years it has been quite clear to me that the stresses and the tensions of daily living are keenly noted by fibrositis sufferers. The target organ where these tensions express themselves is primarily the muscles. The tensions are often quite apparent, and the relationship with muscular symptoms obvious. At times, however, the tensions are not apparent and remain subliminal. That is, the tension stimulus is below the threshold of conscious awareness, having been repeated so often that it is effective on a subconscious level. As a result, chronic concerns and tensions may remain with us, even when we think we feel completely relaxed. Superficially, then, there may seem to be little or no apparent relationship between our tensions and muscular symptoms, while in fact this association exists. Studies at the University of Illinois College of Medicine support the association of stressful life events with fibrositis.[6]

Temperature—and Humidity—Sensitive Symptoms

"I really prefer the hot weather. It eases up my aches and pains. Maybe I ought to move to Arizona."

"Soaking in a hot bath or a hot tub is just great. The worst time for me is when the air-conditioning in the office blows on my neck. I get to aching so much I could scream."

*"The humidity back East was too much for me. I
really prefer the warm dry climate."*

*"I can tell you when it's going to rain. When the ba-
rometer drops, I start aching."*

*"Heat makes me worse. The best thing for my neck is
a cold pack."*

*"The weather? It doesn't make much difference in the
way I feel."*

Fibrositis sufferers differ some with respect to the
effects of heat, cold, and humidity, but about 90 percent of this population is sensitive to changes in temperature.[4] I mention this because I have known a few
people who have moved to warmer climates because of
their aches and pains only to find that they preferred
cooler weather. A large majority by far, however, prefer
both a warm and dry climate. The soothing effects on
muscle pain of warm water in particular have been
known for at least a few thousand years, as shown by
the abundance of spas, the ruins of some over two thousand years old, all over the world.

Anxiety and Depression

*"Perhaps it's because I'm a perfectionist, but I get
awfully anxious about almost everything I do. Everything has to be just right. The house has to be spot-
less, for example. I can't stand messes, and I get
uncomfortable and tense around people who make
messes. I'm always wondering what people think of
me. It makes me anxious and tense, and I just can't
seem to relax if everything isn't just perfect, and of
course it never is. Also, these muscle aches and pains*

make things much worse. I don't know where they come from, and that just makes me more anxious."

"My wife asks me, do I feel depressed? Of course I'm depressed. Who wouldn't be with all these aches and pains. The doctors say they can't find anything wrong with me, and maybe depression is causing some of my troubles. I don't have anything to be depressed about except this damn rheumatism."

For well over one hundred years, many physicians have felt that anxieties and depressions are more common in patients with fibrositis than in the general population. The question has been raised time and again as to whether these emotional components, as well as poor stress tolerance, precede or follow the muscular symptoms, and whether in either event they play a role in the persistence of the muscular symptoms. This has been one of the most controversial topics in the study of fibrositis, and I'll have much more to say about it later in this book.

GENERALIZED AND LOCALIZED FIBROSITIS

The aches and pains of fibrositis can occur in either a generalized or a localized form. This is important to remember, since the two forms differ somewhat as to causes, which I'll discuss later. The generalized fibrositis pain, as the name suggests, affects many muscle groups all over the body, and in a fairly symmetrical fashion. The muscles that are mainly affected are those in the upper arms and legs and those around the neck, shoulders, and hips.

The localized, which is often referred to as a regional pain syndrome, has a tendency to affect only one specific area, and in an asymmetrical fashion. In the United States, a name commonly given this is the *myofascial pain syndrome*.[7] In Australia, when these symptoms are present in an occupational setting requiring certain repeated muscle activities, the physicians researching these problems there have referred to them as *repetitive strain injury (RSI), repetitive strain syndrome (RSS)*, or *occupational overuse syndrome*.[8] Again, the symptoms closely resemble those of more generalized fibrositis except that they are localized to a specific area in the body, such as the head and neck, an arm and shoulder and the associated side of the neck, or either leg and the same side of the low back. Some investigators have, with difficulty, argued that there are specific types of regional pain syndromes. Myofascial pains are, for example, supposedly associated with "trigger points," in which pressure on one point in a muscle causes pain further down the muscle.

All of these variously named localized pain syndromes are actually remarkably similar to one another and are closely related to the more diffuse and generalized form of fibrositis. In an individual patient, the name given to the regional pain syndrome will depend on the training and bias of the person who makes the diagnosis, and, in particular, what country the physician and patient reside in.

The variously named but similar regional muscle pain syndromes are extremely common both in industrial and in nonindustrial settings. They are often found in people who put more strain on one side of the body than the other by various means, and are generally associated with such physical activities as rapid repetitive motions, frequent forceful motions, or with keeping the muscles in a tense position for long periods of time.[9]

FIBROSITIS AND THE TOWER OF BABEL

It seems as though every writer on the subject of chronic muscle aches and pains has decided to give the condition a new name. If you tried to review the medical literature on muscle problems fitting the descriptions of fibrositis and the regional pain syndromes, you would probably feel at some point that you had been trapped on the Tower of Babel.

Fibrositis is the most common term in the English-language medical literature to describe the generalized muscular symptoms we've been discussing. This term was proposed by Sir William Gowers in 1904 in an article in *The British Medical Journal*.[10] Dr. Gowers believed that there was inflammation present in the muscles of patients reporting the symptoms despite the fact that no one had been able to demonstrate it. He said,

> I think we need a designation for inflammation of the fibrous tissue.... We may conveniently follow the analogy of "cellulitis" and term it "fibrositis." ... There is no indication of the formation of "inflammatory products," as we call them, but this is certainly not enough to justify a denial of its inflammatory nature.

The initial response among the medical community to the term *fibrositis* was mixed. Some agreed with Dr.

Gowers, but others were vehemently opposed to "burdening the literature" with such a "meaningless term" or "wastebasket term." Nevertheless, the term *fibrositis* gained general, but not complete, acceptance over the years. Many rheumatologists and other physicians still feel that the name *fibrositis* is a horrendous misnomer and recommend that it be discarded because the name suggests an inflammation that has not been shown to exist.

The only other heavyweight contender for the English-language title is *fibromyalgia*, which began appearing in the medical literature with some regularity in the 1980s. This term combines the words *fibrous*, referring to the fibrous connective tissue in the muscles, and *myalgia*, which simply means muscle aches. *Fibromyalgia*, which denotes fibrous and muscle aches and pains but not inflammation, is slightly more acceptable to the purists.

Those with a more historical bent prefer the use of the older term, *fibrositis*. In hopes of avoiding confusion and controversy, many writers refer to the situation at hand as *fibrositis/fibromyalgia*, *fibrositis (fibromyalgia)*, or *fibromyalgia (fibrositis)*.

In regard to the regional pain syndromes, physicians worldwide have similarly tried to settle on a reasonable name that would gain general acceptance. The name or names they picked often differed according to the country. Many international conferences and congresses have devoted much of their time to standardizing the names, with relatively little success.

Lest you think I am magnifying the semantic problem, let me offer you a partial list in Table 1.1 of the many terms used over the years for conditions comprising the symptoms of fibrositis/fibromyalgia. These

TABLE 1.1
THE MANY NAMES OF FIBROSITIS/FIBROMYALGIA

Anxiety Neurosis	Neurofibrositis
Cervicobrachial Syndrome	Nodular rheumatism
Chronic Nervous Exhaustion	Nonrestorative Sleep Syndrome
Chronic Pain Syndrome	Occupational Cervicobrachial
Chronic Rheumatism	Disorder
Cumulative Trauma Disorders	Occupational Myalgia
Fat Herniations	Occupational Overuse Syndrome
Fibromyositis	Occupational Strain Syndrome
Fibropathic Syndrome	Pain Amplification Syndrome
Functional Myopathy	Piriformis Syndrome
Hypersensitive Areas	Pressure Point Syndrome
Interstitial Myofibrositis	Primary Fibrositis
Lesion Areas	Psychogenic Rheumatism
Low Back Syndrome	Repetitive Strain Injury
Muscle Callus	Repetitive Strain Syndrome
Muscle Gelling	Rheumatic Myitis
Muscle Hardening	Rheumatic Myopathy
Muscular Rheumatism	Rheumatic Myosis
Musculofascial Pain	Rheumatic Pain Modulation
Muskelhärte	Disorder
Muskelschwiele	Scapulocostal Syndrome
Myalgia	Sensitive Areas
Myodysneuria	Sensitive Deposits
Myofascial Pain Dysfunction	Soft Tissue Rheumatism
Myofascial Pain Syndrome	Splanchnic Neurasthenia
Myofasciitis	Tension Fibrositis of the Legs
Myofibrositis	Tension Myalgia
Myogelose	Tension Myalgia of the Pelvic Floor
Myopathia E Labore	Tension Neck
Myositis	Traumatic Myofibrositis
Myospastic Syndrome	Traumatic Neurosis
Nerve Point Syndrome	Trigger Point Syndrome
Neurocirculatory Asthenia	Trigger Zone Syndrome

terms reflect many philosophies of thought; some refer to symptoms, some to the location of the symptoms, others to a hypothesized cause, and still others are purposefully vague.

Consider where the terms in Table 1.1 originate. They come from the many writers interested in finding and documenting causes and treatments for these persistent muscular symptoms. These investigators include family practitioners, internists, rheumatologists, orthopedic surgeons, neurologists, physiatrists (physicians who specialize in physical therapy and rehabilitation), physical therapists, psychiatrists, psychologists, masseurs, and masseuses. This grouping alone suggests that the discomfort of fibrositis sufferers has led them to a wide range of professionals in search of some form of respite from their aching, pain, and stiffness. These professionals were obviously sufficiently impressed with the extent of the problem to give specific names to the disorder and write about it. If you had a compendium of their contributions, you would find that each of the specialties they represent has something very positive to add to the treatment of fibrositis. You would also find that no single medical specialty or discipline today offers the only or the best therapy for fibrositis. Rather, as I have noted, the answer lies in a combination of therapies based on treating an individual with a disorder and not simply treating a disorder alone.

FIBROSITIS, RHEUMATISM, AND ARTHRITIS

I've often been asked if fibrositis is an arthritis or a form of rheumatism. The terminology we use in describing the rheumatic diseases has evolved over several thousand years and is not always used accurately by the general public. Examining a few of these words and clearing up some unnecessary confusion will help us discuss fibrositis with more clarity and pinpoint exactly where fibrositis fits in to the general scheme of the rheumatic diseases.

RHEUMATISM	
Arthritis	Soft Tissue Rheumatism
Osteoarthritis	Fibrositis
Rheumatoid arthritis	Tendinitis
Gout	Bursitis
Metabolic arthritis	Ligamentitis
Infectious arthritis	Capsulitis
Ankylosing spondylitis	Polymyalgia rheumatica
Lupus erythematosus	Giant cell arteritis

When we speak of *rheumatism* we're referring to aches and pains that originate in the bones and joints and their soft supporting tissues, such as muscles, tendons, and ligaments. Rheumatism is in fact a generic term, and as such covers all of our musculoskeletal aches and pains. The term *rheumatismos* was used by Hippocrates in relation to aches and pains about 400 years before the birth of Christ and stems from the Greek concept of disease. The word *rheum* can be translated from the Greek as a flux, flow, or current— movements of various body fluids known as humors that can cause disease. So if you have rheumatism, that means you must have some aches and pains somewhere in your musculoskeletal system. The diagnosis is no more specific than that.

The word *arthritis* also stems from a Greek word, *arthron,* meaning joint. The suffix *itis* means inflammation. There are more than one hundred disorders that can affect our joints. Some are degenerative (osteoarthritis), some inflammatory (rheumatoid arthritis, ankylosing spondylitis), some immunologic (lupus erythematosus), some infectious (gonorrhea), some metabolic (gout), and so on. *Arthritis,* then, means inflammation or damage to a joint. Conversely, if there is no joint problem, there is no arthritis present.

Now for *soft tissue rheumatism.* This means rheumatism affecting the soft tissues of the body that support our bones and joints. These tissues are the muscles, tendons, bursas, ligaments, joint capsules, and fascias. As in arthritis, many different factors—such as trauma, inflammation, and overuse—can cause a variety of conditions: tendinitis, bursitis, ligamentitis, capsulitis, and so on. Polymyalgia rheumatica (PMR) and giant cell *arteritis* (GCA)—not *arthritis*—are also forms of soft tissue rheumatism but are inflammatory diseases of unknown cause.

Fibrositis falls into the category of soft tissue rheumatism and is a type of noninflammatory muscular or nonarticular (not joint) rheumatism. Again, fibrositis is not an arthritis. It is, however, very common and often accompanies all the arthritic and other rheumatic disorders I have listed above. As you might suspect, if fibrositis is present with the other disorders and is not recognized as a distinct condition, the treatment for rheumatic pains will not be fully effective, since the treatment for fibrositis is distinct from that for arthritis. Diagnosing fibrositis as arthritis is far too common.

Fibrositis is a syndrome, as you will find, with distinctive features that set it apart from all other known rheumatic conditions.

CHAPTER

2

Generalized Fibrositis

INTRODUCTION

Medicine has made, and continues to make, truly revolutionary advances in the latter half of this century.
We now have tools that enable us to visualize viruses,
dissect the very essence of body cells, unravel genetic
codes, design potent drugs three-dimensionally on
computers, and use bacteria as little factories to produce hormones such as insulin and cancer-fighting
drugs such as interferon. With computerized tomogra-
phy and magnetic resonance imaging we can detect
disease and distortions of body organs and tissues in a
manner that could only be dreamed of a few decades
ago.

These technological advances have made us more
demanding of "scientific proof"—some form of tangi-

ble and objective evidence of physical abnormality—
before acknowledging that any physical symptoms we
might have really exist. Physicians, certainly, are not
immune to putting a great emphasis on what we might
call this "seeing is believing" philosophy.

If you have fibrositis, however, you may have already discovered that with all our space-age technology we still have no simple fibrositis test or any
combination of laboratory tests or X rays that allows us
to see or prove the diagnosis. In fact, if your physician
depends primarily on the "seeing is believing" approach, he or she may find it hard to accept that there
is such a thing as fibrositis, or that your symptoms are
of much significance. You'll know if you have encoun-
tered it that this skeptical attitude can only add to the
misery of your very real aches and pains.

This chapter is designed to dispel any such skepti-
cism on the part of doubting physicians. In addition, it
will provide the fibrositis sufferer, who knows the
symptoms are real, with both reassurance and the insights needed to embark on an effective treatment program.

First, we'll focus on some of the statistics regarding
the incidence of fibrositis, followed by some historical
aspects on just how the concept of fibrositis evolved.
This will be much more than an academic exercise,
since many of the clues as to what we know about the
causes and best avenues of treatment of fibrositis are to
be found in statistics and historical data.

A FEW STATISTICS

Before the 1970s, it was next to impossible to get any
clear idea of how many people in the general popula-

tion were suffering from generalized fibrositis. The only information we had at that time came from military sources, and as you might expect, it applied almost entirely to men. Statistics on rheumatic illnesses among the four million soldiers in the American Army of World War I showed that fibrositis (muscular rheumatism, as they called it then) was the third most common rheumatic disorder affecting military personnel, involving 15 percent of all soldiers with muscle or joint problems. The number of cases of fibrositis was surpassed only by rheumatoid arthritis and degenerative joint disease.[1]

Let's compare that finding with those from studies done in the British and American armed forces during World War II. These data, of course, also related mainly to men. Reports of illnesses in the British Expeditionary Forces showed that 26 percent of all medical patients admitted to several general hospitals had rheumatic disorders, and 70 percent of those patients were diagnosed as having fibrositis.[1] By extrapolation, we can determine that a startling 18 percent of all the armed forces medical patients admitted to these general hospitals in Britain had a diagnosis of fibrositis! A study done during World War II in the United States revealed that of 5,200 patients admitted to the Army Rheumatic Center at Ashburn General Hospital, 3 percent were diagnosed with fibrositis, but 17 percent were diagnosed with psychogenic rheumatism.[1]

It is obvious that severe muscular pains were common enough in British and American soldiers during World War II, accounting for approximately 20 percent of the rheumatic illnesses requiring hospitalization. But the studies also revealed another pertinent fact: The condition that physicians in Britain were naming fibrositis was being called psychogenic rheumatism in the United States. Psychogenic rheumatism is defined as aches and pains that are solely in the mind, with no basis in reality. As I noted before, the diagnosis of these aches and pains by physicians in the American armed forces in World War I was muscular rheumatism, but in World War II the same condition was given the name psychogenic rheumatism. This change in nomenclature suggests that American, but not British, army physicians underwent a marked shift regarding the cause of this disorder—in their minds it was transformed from a vague, nonspecific illness to a psychologically derived condition.

One of the truly obvious facts confirmed by these studies, however, was left unsaid in the reports: In times of great danger, when personal stress is extreme, the incidence of muscle aches and pains in men increases drastically. If such statistics regarding muscle aches and pains had been reported in peacetime, they would most likely have been called a great epidemic among young men.

A report from the Mayo Clinic in 1977 revealed some fascinating figures regarding the incidence of fibrositis.[2] Dr. Henry H. Stonnington reviewed the diagnoses of almost 21,000 patients who had been treated in the previous year at Mayo's Department of Physical Medicine and Rehabilitation. These patients were being treated for a variety of disorders, such as arthritis and other rheumatic diseases, injuries, and stroke. He found that 11 percent of these patients had a diagnosis of fibrositis. Dr. Stonnington wrote,

Although these syndromes are vague, they are probably one of the most important causes of morbidity [the rate of disease] and work absenteeism. . . . This is therefore an immense problem in health care which not only taxes the ingenuity of the physician . . . but also the endurance of the patient.

In 1982 and 1983, studies done in the United States and Mexico on patients seen in rheumatologists' offices showed that fibrositis was the diagnosis in 15 percent and 16 percent, respectively.[3,4] Yet another study in 1983 showed that fibrositis was found in 5.7 percent of all patients in a university's general medical clinic.[5]

A report from the rheumatology clinic of the Peoria School of Medicine in 1982 noted that in 1979, 20 percent of their patients seen were diagnosed with fibrositis. In the patients under age fifty, fibrositis was diagnosed in an astounding 33 percent![6]

On a recent trip to Budapest, my wife and I were privileged to visit the Hungarian National Institute of Rheumatology and Physiotherapy. We took full advantage of the wonderfully relaxing hot baths and massages offered by the many cosmopolitan spas in the city. The guests in these spas had journeyed from the various countries of the Eastern and Western bloc nations to get relief from their muscle and joint aches. Rheumatic pains, it seems, have no political boundaries! My host was Dr. Geza Balint, a fine clinical rheumatologist and editor-in-chief of the journal *Hungarian Rheumatology*. I asked him what the most common rheumatic problems were in Hungary. His answer, which he had just published: degenerative arthritis, soft tissue rheumatism, and fibrositis.[7]

In an article in the *Journal of the American Medical Association* in 1987, Dr. Donald L. Goldenberg stated, "Fibromyalgia [fibrositis] is a common condition.

There are 3 to 6 million patients diagnosed with this condition in the United States, but this probably grossly underestimates the prevalence of this syndrome."[8]

The conclusion we can draw from these findings is surely obvious. Fibrositis has emerged not only as one of the leading causes of the chronic pains of all forms of rheumatism but as the major cause of pathological muscular aches, pains, and stiffness in every country where it has been investigated.

FIBROSITIS: EVOLUTION OF A CONCEPT

The early search for relief from generalized muscle aches and pains centered around attempts to find a single cause to account for all the symptoms. A great hope of medical researchers investigating any medical disorder has always been to find a single cause for the illness, then discover a specific treatment to eradicate the cause, thereby effecting a cure of the disease. This is a wonderful approach when it works. And it has worked for example with syphilis and pneumococcal pneumonia. Both these diseases are due to bacteria and can be cured by treatment with penicillin. Another example is scurvy, the famous plague of sailors on long sea voyages. It is caused by a deficiency in vitamin C and can be prevented and cured simply by eating foods containing vitamin C or taking the vitamin alone.

Many researchers and practitioners in many different specialties have tried to apply the find-the-cause/find-the-cure approach to fibrositis, with singularly frustrating results. It is just these failures to produce the hoped-for cure that have forced us to think in

broader terms, and fortunately this shift in conscious-
ness has yielded much more satisfactory results. We
can learn a lot about fibrositis by exploring the many
paths taken by the pioneers in this effort. The recent
resurgence of interest in fibrositis, I might add, has
spawned many excellent and exhaustive historical re-
views of mankind's search for relief of muscle aches
and pains.[9–13]

Surprisingly, it wasn't until the seventeen hundreds
that physicians began to distinguish between stiffness
and pains originating in the joints (arthritis) and those
coming from the muscles (muscular rheumatism). During the mid- and latter 1800s, most of the medical literature on muscular rheumatism was produced by the
German physicians and Scandinavian masseurs. In the
early 1900s the British began writing extensively on
the subject, and in the 1930s the Americans belatedly
joined the effort.

Nodules in the Muscles

The first descriptions of muscular rheumatism centered
on the results of physical examination and gave birth to
a controversy that took more than a hundred years to
resolve. The spark that ignited the debate was the finding of painful hard places in the muscles. Dr. Robert
Froriep, a German physician, is given credit (or dis-
credit, depending on whom you read) for first reporting
these in 1843. He called them *Muskelschwiele* (muscle
callus); later, other German physicians called them *My-
ogelose* (muscle gelling) or *Muskelhärte* (muscle
hardening). The investigators stressed that practitioners needed special training to detect these muscle abnormalities. Even with such training, however, not

everyone could feel the precise phenomena described. Soon a whole array of abnormalities had been documented. Some felt to the doctors like grains of sand, some like apple seeds, and some like pea-sized nodules as hard as bone. Still others were described as hardening of larger portions of muscles, soft tender spots, or tender bands of tissue. Such findings came to be considered essential to the diagnosis of muscular rheumatism. Whatever was truly being felt, the general impression was that these findings were abnormal and were the source of all the aches, pains, and stiffness.

Arguments raged as to the cause of these abnormal muscle findings. In Germany and Scandinavia, the cause of the nodules was generally held to be exudate —fluids seeping into the abnormal areas. In France and Britain, however, the problem was thought to be due to some type of irritation or inflammation in the muscles. The nodules were treated with heat, cold, massage, electric currents, injections of anesthetics, and even excisions. These treatments gave temporary, little, or no relief. Many of the nodules, which were later called *fibrositic nodules* in the English literature, were excised and examined carefully under the microscope. Almost everyone's findings about them were consistent: They looked like absolutely *normal* muscle tissue.

Many physicians began to doubt that the nodules (which certainly can be felt in many people, both those who do and those who do not suffer from fibrositis) had any pathological significance at all and to consider them normal findings. In 1936, Dr. Philip Hench, an American physician, wrote that "some writers make much of the nodules of fibrositis, which to others are only accessible to the finger of faith." [14]

Masseurs and masseuses, however (particularly the Scandinavians and the Dutch), considered these nodules very important. They offered massage as the major therapy, which did afford patients some temporary relief. A German physician, Dr. A. Müller, was quoted in 1912 as saying that "the existence of fiber hardenings is denied by most internists and surgeons while masseurs look for and successfully treat them."

Despite all the discussions and disputes, physicians, masseurs, masseuses, and patients all agreed on one thing: Whatever it was called, wherever it was coming from, and whatever helped or didn't help it, a large percentage of the population all over western Europe, Scandinavia, and the British Isles was suffering from these very significant and real muscle aches, pains, and stiffness.

Trigger Points

By the 1930s and 1940s, the talk about nodules had pretty much died down and attention turned toward the importance of trigger points (less commonly called trigger zones) in the muscles. A trigger point was considered to be an area in which applied pressure would cause a referral or triggering of pain elsewhere. An example would be if you pressed a finger on the muscles of your left shoulder blade and the pain was felt radiating into your left shoulder and down your left arm. Many writers described trigger points in various parts of the body; these points could cause pain to shoot down into the shoulders, arms, back, buttocks, and legs. Trigger points were considered an important feature of fibrositis, particularly the localized form of fibrositis known as the myofascial pain syndrome (described in Chapters 1 and 4). The points themselves became sites of treatment and were treated by various methods: primarily by injection with a local anesthetic but also with a spray of a rapidly evaporating coolant such as ethyl chloride or fluoromethane followed by stretching of the involved muscles. Two major proponents of these concepts and treatments, which gained a modest acceptance among rheumatologists, were Dr. Janet Travell and Dr. David G. Simons.[15,16]

Tender Points

An often quoted paper by Dr. Hugh A. Smythe and Dr. Harvey Moldofsky in 1977 discussed two important contributions to the understanding and defining of the fibrositis syndrome.[17] The first was the concept of *tender points*. The second involved evidence suggesting a specific disturbance in sleep that produced or exaggerated the pain. In regard to the first concept, these doctors believed that fibrositis sufferers demonstrated exaggerated muscle tenderness in fairly precise sites. "The reproducibility of these sites across ethnic and cultural boundaries is striking," the doctors wrote. Many other physicians agreed that tender points in specific areas constituted an important finding. In 1986, in discussing the findings of physicians over the past century and a half, Dr. Smythe wrote, "The tenderness described by all these authors is consistent with modern observations, but the indurations [the nodules and hardenings] present more difficulties."

What has become apparent over the years is that the *only* consistent finding on physical examination of the patient with fibrositis is the presence of these many tender spots in specific muscle sites. Many patients do have fibrositic nodules and others have trigger points, but these phenomena are not found in everyone who has fibrositis.

Sleep Disturbances

In the morning when someone asked how she had slept, she replied, "Oh, just wretchedly! I didn't close my eyes once, the whole night through. God knows what was in that bed; but it was something hard, and I am black and blue all over."

Now they knew that she was a real princess, since she had felt the pea that was lying on the bedstead through twenty mattresses and twenty eiderdown quilts. Only a real princess could be so sensitive.

"The Princess and the Pea"
—Hans Christian Andersen[18]

As the translator of this story says, "The fairy tale and the folk tale take place in the real world." The fairy tale "The Princess and the Pea" was first published in 1835, just eight years before Dr. Froriep described his *Muskelschwiele*, or muscle calluses, thereby ushering in the scientific study of muscle aching. As we will see, this was 140 years before any scientific evidence was published linking sleep disturbances with muscle aches and pains, but the connection did emerge.[19]

A very common complaint found among generalized fibrositis sufferers is "poor sleep," or in more technical terms "nonrestorative sleep"—sleep that doesn't seem to restore one's energy and is associated with morning fatigue and worsening of the aches, pains, and stiffness. Certain investigators began to ask whether fibrositis sufferers really do sleep poorly and if so whether the abnormal sleep is a *result* or a *cause* of the fibrositis. Since it is possible to measure some aspects of sleep by looking at the electrical discharges from the brain, these investigators concluded that observing what happens in fibrositis sufferers in sleep could be a way of unearthing the kind of objective findings notably absent in the fibrositis research.

Understanding the thrust of this branch of research requires a little basic knowledge in sleep physiology and its measurement. The study of sleep was revolutionized with a report in 1953 of the discovery of rapid eye movement (REM) sleep. This is a period of sleep during which rapid eye movements occur, and it is considered to be a normal but lighter form of sleep than non-REM sleep, which is characterized by a lack of rapid eye movements. By examining the electrical impulses coming from the brain during sleep with the help of an instrument called an electroencephalogram, researchers divided non-REM sleep into four states. Under their system, stage 4 was deemed the deepest sleep.

Normally, periods of REM and non-REM sleep alternate. A healthy adult will spend about 20 to 25 percent of sleep in REM sleep and the rest in non-REM sleep. The usual sequence is first to spend about 70 to 90 minutes in non-REM sleep, then 5 to 10 minutes in REM sleep. As sleep continues, there are more cycles of non-REM and REM sleep, with the length of time of non-REM sleep shrinking and that of REM sleep lengthening.

In 1975 Dr. Harvey Moldofsky and his colleagues reported finding evidence to suggest that fibrositis sufferers have a disturbance in their non-REM sleep.[19] With fibrositis, the electroencephalogram during non-REM sleep showed the random appearance of what are called alpha-delta waves—a situation not found in normal sleep. Some of the subjects in this study failed to experience the deeper sleep of stages 3 or 4. Interestingly, these researchers wrote, "All subjects reported that major stressful life situations had occurred at the time of onset of their sleep disturbance and musculoskeletal and mood symptoms. Typically, they had experienced a frightening but minor automobile or industrial accident or they had become helplessly involved in insoluble domestic difficulties."

Further, in another study in 1977, Moldofsky and his co-workers interrupted stage 4 non-REM sleep in a group of healthy volunteer university students who had not previously had the symptoms of fibrositis.[20] They saw the appearance of these alpha-delta waves along with complaints of muscle aches, pains, and stiffness in these subjects. Of interest was that those subjects in the group deprived of stage 4 non-REM sleep who were very physically fit and ran three to seven miles a day did not develop the muscular complaints. Further studies confirmed these findings, and suggested that many fibrositis sufferers are trapped in a cycle of emotional distress, nonrestorative sleep, and musculoskeletal pain and fatigue.[21] These studies also suggested—and here we have an important point in regard to treatment—that good physical conditioning had a protective effect against the symptoms of muscle pain and fatigue.

Perhaps here, then, with the evidence on sleep disturbances and brain wave changes and exercise, we have our first truly objective finding to help us explain exactly what is happening in fibrositis.

Immunology and Fibrositis

The search for an objective cause or causes for fibrositis led some investigators to consider the possibility that immunologic abnormalities may be involved in generalized fibrositis.

The rapidly advancing science of immunology initially concerned itself with the study of the mechanisms by which we resist and overcome disease and infection that enters our bodies. This was followed by an expansion into the study of the reaction of the body against any type of foreign substance (*antigen*) that invaded or was introduced into body tissues and stimulated an immune response. In the case of the various rheumatic diseases, particular interest has centered on what is yet another stage—the *autoimmune diseases*. Autoimmunity is the state of being immune, or allergic, to our own body tissues. In short, in autoimmune diseases, the body mistakenly considers certain of its own tissues and organs as foreign substances and begins to attack them.

A small percentage of patients with generalized fibrositis symptoms have laboratory findings suggestive of an autoimmune problem, such as a mildly elevated rheumatoid factor or antinuclear antibody in the blood (which I will discuss in more detail later) or Raynaud's phenomenon. Raynaud's phenomenon is a condition where the small blood vessels of the fingers and toes go into spasm on exposure to cold, causing blanching and pain in the affected digits. A positive rheumatoid factor or antinuclear antibody suggesting the presence of an autoimmune disease can be found in many rheumatic diseases; however, it can also be found in at least 5 percent of the perfectly normal population. The situation with Raynăud's phenomenon is similar: It can occur in many otherwise perfectly normal individuals.

Although some patients with autoimmune diseases such as lupus erythematosus may have symptoms vaguely suggestive of fibrositis as the first sign of their disease, the true picture usually clarifies itself rapidly. The findings to date, however, indicate that people suffering with fibrositis symptoms do not develop an autoimmune disease. The issue of whether or not autoimmunity may play a role in the symptoms of some people with fibrositic muscle aches and pains has not yet been finally resolved.[22]

Genetics and Fibrositis

I've often been asked whether fibrositis can be inherited, or in other words, whether genetic factors play any part in this disorder. At this time, there is no evidence to allow a definite answer one way or the other. However, my personal feeling, based on reviewing family histories from many people, is that genetics and heredity do play some role in determining susceptibility to fibrositis, although it is difficult to separate out the factor of learned behavior.

DO YOU HAVE FIBROSITIS?

Until now I've given you a lot of background on what fibrositis is, what it isn't, and how the concept evolved. At this point we'll set a goal—that of helping you determine more accurately whether or not your muscle aches and pains might be caused by fibrositis. Again,

please note that I use the word *might*, since this book
is meant to be a guide to your thinking and not a manual on self-diagnosis. If you have the symptoms of
chronic muscle aches, pains, and stiffness, you should
first consult with your physician. You may suspect your
problem is fibrositis, and you may be right—or you may
be wrong. A myriad of other disorders can cause symptoms that superficially resemble those of fibrositis.
Most of these disorders are less common than fibrositis,
but some can be more serious and will require different types of therapy. On the other hand, even if you
have other medical problems, don't forget that your
muscle symptoms may still be due to fibrositis.

To achieve our goal, we're going to follow the path
that rheumatologists are now using to make a diagnosis
of generalized fibrositis. On this path you will learn of
the various criteria that define fibrositis. For many
years, I might add, there were no clear-cut guidelines
for physicians that would help to make the diagnosis of
fibrositis. It wasn't until the 1970s that rheumatologists
themselves felt confident enough to propose the criteria that could be used for the diagnosis of generalized
fibrositis. It then took another decade for the criteria to
be refined or "tuned up" and be generally accepted
within the medical community.[23]

The three major criteria that rheumatologists and
other physicians can now use to aid in the diagnosis of
fibrositis are

- the presence of muscle aches, pains, or stiffness for at least three months (the history)
- the presence of tender points in specific muscular anatomic sites (the physical examination)

- the absence of any other underlying disorder that would fully explain the muscular symptoms (evaluation of the laboratory studies and X rays)

The History

During my medical training, one of my professors somewhat facetiously identified the three most important points in making a diagnosis as the history, the history, and the history. His point was that in many illnesses, doctors can find the major clues to the diagnosis by listening carefully as the patient relates his or her complaints instead of relying primarily on the results of laboratory and X-ray studies. This advice is particularly pertinent in the case of fibrositis. A good history is not simply important, it is absolutely essential in determining the presence or absence of fibrositis. The history that the physician obtains, and that you yourself must think about, should include not only the muscular symptoms but a review of everything else that seems to be going wrong in your body.

The sine qua non of fibrositis is a history of muscle aches, pains, or stiffness. The muscles most frequently involved are those of the neck, the shoulder, and hip areas. The discomfort can be widespread, however, also involving your low back and all four extremities. Most people tend to equate muscular aches and pains with stiffness, but the two kinds of sensations are distinct. It is possible to have aches and pains with no stiffness, and stiffness with no aches and pains.

Let's consider the pain first. Generalized fibrositis sufferers variously describe their pain as aching, burning, deep, sharp, dull, knifelike, needlelike, steady, intermittent, fatiguing, gnawing, spasmodic, throbbing, tingling, numbing, cramping—the list goes on and on. Often even the skin is involved, having sensations of tingling, burning, numbness, heat, and cold. The descriptions that sufferers offer of their pains and discomforts suggest either that fibrositis has various effects in different people or that similar muscle discomforts are described differently by different people. After questioning many hundreds of fibrositis patients, I hold with the latter.

As to muscle stiffness, it simply means that it's hard to move your muscles—the muscles seem tight, not free or limber. It is as if the muscles have jelled. This stiffness is generally most apparent after a period of inactivity, such as the morning after a night's sleep or after sitting for a long period of time. The stiffness is gradually relieved by activity or by the application of heat. After a period of resting or exposure to cold, however, the stiffness will generally return.

Many physicians are hesitant to make a diagnosis of generalized fibrositis unless the muscle pains have been present for at least three months. The three-month criterion excludes transient illnesses, such as the flu or other viral or bacterial infections that can cause such aching and pain. These other illnesses rarely last for three months. You can suspect fibrositis and treat it at any time, but you and your doctor should be especially alert for other problems during the three-month waiting period. On the other hand, there is no restriction on how long the symptoms can be present. I recently saw a sixty-four-year-old woman in my office who had suffered the typical symptoms of fibrositis—undiagnosed—for forty-four years!

Fatigue, as noted in the first chapter, is high on the

list of other symptoms that accompany the fibrositis syndrome, but it is not part of the criteria necessary for making the diagnosis. The fatigue can be mild or it can be severe enough to interfere with your daily activities. Unlike the normal fatigue that comes after a long day's work or after considerable muscle exertion, that of fibrositis is present on awakening and generally persists throughout the day. Resting usually doesn't help very much. Many fibrositis sufferers find that the fatigue concerns them as much as muscular aches and pains. Often, these very complaints of severe fatigue in undiagnosed fibrositis raise the specter of some serious underlying disease, prompting the physician to order a battery of expensive laboratory studies and X rays that ultimately prove to be normal.

To reiterate factors that can contribute to the diagnosis, in addition to the fatigue, irritable bowel syndrome, headaches, sleep disturbances, cold intolerance, feelings of joint swellings, tension, anxiety, and depression all should cause your doctor to consider the fibrositis diagnosis.

A Case Study

Almost universally, physicians experienced in diagnosing and treating generalized fibrositis agree that a history of certain characteristics and life-styles is found more frequently among patients with this condition than in the general population. I am not saying that all fibrositis sufferers have all these traits, but to a remarkable degree threads of similarity run through the fabric of the lives of people, both male and female, who have fibrositis. Let's put the statistics and histories together to create a profile, a character sketch and history of the typical person plagued by the constant muscle aches and pains that we call fibrositis. I'll call our subject Carolyn:

Carolyn is a woman in her early thirties who has had fibrositis symptoms for a few years. She has sought help from various physicians and chiropractors and has been told on different occasions that nothing could be found, that she may have arthritis, or that her spine might be out of alignment. None of her many laboratory and X-ray studies have shown any significant abnormalities that would explain her pains. Over the years, a dozen or more therapies have been prescribed, including anti-inflammatory agents, muscle relaxants, and spinal manipulation—all to no avail. Carolyn has tried various diets and many bottles of vitamins and minerals with no significant relief. And she's worried. She knows the pains are not in her mind, and she is fearful that she has a disease that at some point will manifest itself in deformity and disability.

Carolyn's health has otherwise been very good, and her physical appearance does not suggest that she is chronically ill. When she was younger she was very active physically, but she's too busy now to exercise as much as she used to, and she feels a bit out of condition. She has trouble sleeping at times, and when she awakens at night she often thinks of what has to be done the next day.

When she was in school, her grades were above average, and she studied various subjects beyond her high school days. She looks on education as an ongoing process that does not end at the twelfth grade. She has always considered herself a good worker of above average reliability, and those who know her confirm this view.

Carolyn is sensitive to criticism, and takes any complaint very personally. She avoids controversy whenever she can, since it adds to the stress she feels and makes her very uncomfortable.

Carolyn is compulsive—not in the psychiatric sense of being irrationally impulsive but in the sense of having a strong driving force behind her accomplishments. In short, she puts a high value on success. She has been called a perfectionist at times and has trouble living up to the standards she sets for herself.

Perhaps owing to her sensitivities and expectations of herself and others, Carolyn finds it difficult to relax. Nothing ever seems to be fully done; she always has the feeling that something remains unfinished that demands her attention. Rather than waiting for someone to do his or her job, she will often do it herself.

As to Carolyn's relationships with others, she has been described by her friends and coworkers as caring and sensitive to people's needs. She does expect others to live up to her own standards, however, and this inevitably brings her into conflicts and stressful situations. When she is not working full-time, she can be found serving as a volunteer in various organizations. She is well dressed but not a clotheshorse; she doesn't allow her clothes to draw undue attention to herself.

Her tensions are often associated with anxieties about what people think of her and what the future will bring. She has times of relaxation, but they are often disrupted by thoughts of something that has been neglected. At times she feels somewhat de-

pressed, and then she cries easily. She has many internal conflicts about her relationships with others, and it seems to her she has a harder time resolving these conflicts than do those around her.

Carolyn's stresses and tensions are magnified by the constant aching, pain, and stiffness in her muscles. Together these symptoms are very fatiguing and seem an added burden to bear. She feels that if she could just get rid of these worrisome symptoms, the other stresses and tensions would seem minor and her life would be easier.

The characteristics cited in the sketch of Carolyn are found among fibrositis sufferers in every country in which the disease has been studied. I present this composite for specific reasons: to heighten your awareness of who is likely to get fibrositis and to suggest why an understanding of people's characteristics and lifestyles can help us in diagnosing and treating this disorder.

The Physical Examination

The second criterion necessary for a diagnosis of fibrositis is the finding of tender points in specific muscular anatomic sites. This requires a careful physical examination on the part of your physician. If you have fibrositis, you can easily find many of these points yourself by pressing a finger on the involved muscles. Figure 2.1 shows the most common tender points.

Certainly if enough pressure is applied, everyone will feel discomfort and pain in these areas. The amount of pressure we're talking about, however, is moderate fingertip pressure that does not cause pain in

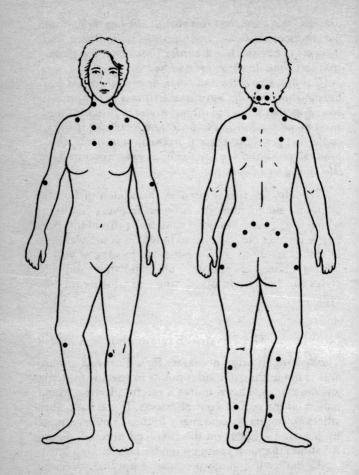

FIGURE 2.1: *Frequently Found
Tender Points in Fibrositis*

the average individual but causes quite a bit of pain
and discomfort in the fibrositis patient. Attempts have
been made to standardize the pressure applied through
the use of an instrument called a dolorimeter, which
gives a measure of the pressure applied. The results
have not been very reproducible, and most physicians
rely on their clinical experience obtained from examining many patients. If you've had any personal experience with acupressure or the Japanese massage
technique known as shiatsu (meaning "finger pressure"), you'll find that both of these treatment techniques use pressures far beyond that needed to cause
pain in a fibrositis patient.

How many tender points are required for a diagnosis of fibrositis? Rheumatologists are not completely
unanimous regarding the answer to this question. The
number suggested ranges from four to fourteen or more
for generalized fibrositis, and less for the regional pain
syndrome. Studies have shown that 96 percent of patients with generalized fibrositis have more than four
tender points, and 94 percent have more than seven.[23]

Although not part of the recent criteria for the diagnosis of fibrositis, very tender "knots" or lumps in
your muscles that were formerly given the name *fibrositic nodules* may be part of your set of symptoms. You
may feel the tender muscular bands or ridges previously described, particularly in the muscles of the neck
and upper back between the shoulders. Massaging
these areas and other tender points will cause pain initially, but many people find that this discomfort is followed by a period of decreased pain and muscle
relaxation.

Laboratory Studies: Help or Hindrance?

The third criterion for the diagnosis of fibrositis is the differential diagnosis—the "what else can it be?" part. An essential ingredient here is a careful choice of laboratory studies and accurate interpretation of results. Your physician will probably review the findings with you, so a little background will be of help to you in understanding their meaning. If your physician suspects that you may have some musculoskeletal or rheumatologic disorder, he or she will order some specific laboratory studies to help either confirm or rule out the possibility of various illnesses. If these tests are interpreted correctly, they can aid in the diagnosis considerably. If they are interpreted incorrectly, they can be a terrible hindrance to reaching the proper conclusion. In short, laboratory tests can be a blessing or a curse.

There are thousands of different blood tests being performed daily in clinical laboratories in this country, many of which are useful in the diagnosis of the disorders we call rheumatism. Three tests are most frequently performed, and these are considered the basic, or screening, ones. These tests are the *erythrocyte sedimentation rate (ESR), rheumatoid test for rheumatoid arthritis (RA test)*, and the *antinuclear antibody test (ANA)*.

The ESR is usually referred to as the "sed rate." The test measures how far blood cells and platelets fall, or sediment, in one hour. To perform this test the physician or technician draws blood from a vein with a needle and syringe into a tube containing a tiny amount of anticoagulant that prevents the blood from clotting. A small amount of this blood is transferred to a long, thin, calibrated glass tube and allowed to stand vertically for one hour. During this time the formed elements of the blood—the red and white cells and the platelets—slowly settle, or sediment. The result is a tube with packed cells in the lower part, sharply demarcated from the clear plasma in the top part. The actual sedimentation rate is the number of millimeters (mm) that these formed elements have fallen in one hour, which is read by measuring the length of the clear plasma. Some laboratories use a tube 100mm long (the Wintrobe method), while others use a tube 150mm long (the Westergren method). The Westergren method gives higher readings and is ordered by most physicians, since it is more sensitive to changes in the blood than is the Wintrobe method.

In young healthy people, the ESR is 20mm or less. As we get older, the normal value can be as high as 30 or 40mm. Readings above this are considered to be abnormal and can be due to many causes. High ESR readings are not specific for any disease and can be found in a host of disorders, including rheumatoid arthritis, lupus erythematosus, polymyalgia rheumatica, giant cell arteritis, inflammation or infection from almost any source, and the presence of abnormally large amounts of proteins in the blood. A high reading—and this is an important point—can also occur in people who are perfectly healthy!

Now for the help or hindrance. In a patient with generalized fibrositis symptoms, the ESR is expected to be normal. If the physician does not consider a diagnosis of fibrositis, a normal test would be one piece of evidence suggesting to him that all is well, "You don't have arthritis," and perhaps you are simply imagining your aches and pains. If the ESR is high (again, a situation that can happen by chance in fibrositis patients as well as perfectly normal and healthy people), he or she might suspect arthritis or a host of other conditions and continue along with a seemingly endless battery of expensive laboratory studies. I have seen both of these scenarios occur when the diagnosis of fibrositis was not considered.

The RA and ANA tests can cause even bigger problems than the ESR. The rheumatoid factor, or RA test, is a measure of a protein found in the blood of people who have rheumatoid arthritis. What is often not appreciated, however, is the fact that this protein can be found in smaller amounts in the blood of people who are perfectly normal. The designers of the test, who were well aware of this, set the sensitivity of the test at a point where it would be positive in most cases of rheumatoid arthritis and negative in people who did not have rheumatoid arthritis. The end result is that about 85 percent of people with rheumatoid arthritis have a positive test, but about 5 percent of normal adults also have a positive test. If they had set the sensitivity at a point where it would detect almost all of the people with rheumatoid arthritis, then a very high percentage of normal people would also be positive, making the test almost totally useless.

Now suppose you have muscle aches and pains—but no rheumatoid arthritis—and you are one of the 5 percent of people with a positive RA test. If your physician puts undue reliance on this test, you will be given an unwarranted diagnosis of rheumatoid arthritis and will be treated as having it. I can recall many fibrositis sufferers who were RA positive and erroneously diagnosed as having rheumatoid arthritis, despite the complete absence of joint swelling, on the basis of this test alone. Many of these patients were given a variety of medications with potentially serious side effects but that had little or no effect on the muscle aches and pains.

The antinuclear antibody or ANA test involves the identical pitfalls. This is an immunologic test designed to detect an antibody in the blood that reacts with the various constituents in the nucleus of certain body cells. A fluorescent compound used in the test allows the laboratory technician to see, under a microscope, exactly where the antibody is reacting with the cell nucleus. It forms certain patterns that are reported as speckled, homogeneous, and the like, and each pattern tends to be loosely associated with specific diseases. A positive ANA test is found in almost all patients who have systemic lupus erythematosus and some who have rheumatoid arthritis, Sjögrens syndrome, scleroderma, certain types of hepatitis, and so on. Again, unfortunately for exacting diagnostic purposes, the ANA test is positive in about 5 percent of normal adults who have no evidence of any underlying disease. Also, the older you are, the more likely you are to have a positive ANA.

I do not want to minimize the importance of laboratory studies. Given the vagaries of these tests, however, a diagnosis of any rheumatic condition depends on a careful integration of the results of the history, physical examination, and laboratory studies, not on any one of these approaches alone.

If your diagnosis is based primarily on the results of laboratory studies, and you suspect that it is incompatible with what's going on in your body, I strongly suggest that you get a second opinion from a rheumatologist skilled in interpreting all of this information.

X-ray Studies: Help or Hindrance?

X-ray studies in the diagnosis of fibrositis not only have all the problems of laboratory studies, but fall into a special category of their own. First, fibrositis itself produces no abnormal X-ray findings, even in the severest of cases. Again, generalized fibrositis is a disorder of the soft tissues, not of the bones or joints. The use of X-ray studies is helpful only in determining if some other disorder, such as arthritis or a bone condition such as osteoporosis, is present. Since fibrositis can occur in the presence of these and many other conditions, your physician may feel it is appropriate to order specific X rays to help in the total diagnosis. The pitfalls and hindrances come in when arthritis is present and the symptoms of fibrositis are attributed solely to the joint disease. Arthritis and fibrositis differ significantly, and treatment of the former will generally give little or no relief to the latter. I have seen many patients with fibrositis and insignificant X-ray changes of arthritis in the neck who were treated with anti-inflammatory drugs and neck traction for long periods of time, only to have their symptoms get worse.

Also, although the dangers of radiation from a few bone and joint X rays is practically nil, such studies should be ordered by your physician only when absolutely necessary. Keeping a possible diagnosis of fibrositis in mind will obviate the need for many X rays and the consequent exposure to radiation. In addition, avoiding unnecessary X rays has some obvious cost benefits.

THE DIFFERENTIAL DIAGNOSIS

The muscle aches and pains of generalized fibrositis are usually so unique in their presentation that they can be correctly suspected in most cases of being the major culprit by history alone. A few disorders, however, can mimic some of the features of generalized fibrositis and are both important and common enough to bear mention, since they require different treatments than fibrositis does.[24] These disorders include polymyalgia rheumatica, giant cell arteritis, multiple areas of tendinitis, and hypothyroidism.

Also relevant are two controversial conditions currently receiving a lot of attention in the press: Epstein-Barr Virus infection and the chronic yeast infection known as candidiasis or moniliasis. Both of these conditions allegedly cause chronic muscle pains; however, whether they actually do remains highly questionable.

Polymyalgia Rheumatica and Giant Cell Arteritis

Polymyalgia rheumatica (PMR) and giant cell arteritis (GCA) are two closely related disorders of unknown cause that affect men and women over fifty years of age. Unlike fibrositis, and this is important, they are rarely seen in younger people.

The term *polymyalgia rheumatica* simply means many muscle aches of a rheumatic nature—not a very specific term. It is characterized by muscle stiffness and pain in the neck, shoulder, and hip areas that is generally more pronounced than that occurring in fibrositis. The stiffness can be so severe that the afflicted person can barely get out of bed in the morning and get dressed. Sitting or any type of inactivity causes the muscles to become even stiffer. The pain can vary and is usually most intense during the night. Sleep is problematical owing to the combination of pain and difficulty turning over in bed.

As in fibrositis, the X-ray studies are normal. The primary diagnostic tool for this condition is the sedimentation rate (ESR) that I have described. The ESR is almost always above 50mm per hour when done by the Westergren method, and can be as high as 100 or more mm per hour. In this situation, the ESR is of great help in the differential diagnosis. Polymyalgia rheumatica, as you can imagine, can easily be mistaken for fibrositis, and vice versa.

Giant cell arteritis (also known as temporal arteritis and cranial arteritis) can cause a wide range of symptoms. It may begin with symptoms identical to polymyalgia rheumatica. The disease gets its name from the fact that under the microscope one can see definite inflammation of various arteries (called arteritis) and can find large cells with many nuclei in the vessel wall. Such inflammation can cause damage to many tissues, the most feared being damage to the blood vessels supplying the retina of the eye, from which blindness results. As in polymyalgia rheumatica (but unlike fibrositis), the ESR is usually elevated and frequently over 100mm per hour.

Both polymyalgia rheumatica and giant cell arteritis can be treated very effectively with a cortisone derivative. Prednisone is most commonly prescribed and can result in dramatic improvement. Prednisone, however, has little or no effect on the symptoms of generalized fibrositis.

Tendinitis

Tendinitis (or tendonitis), as the name suggests, is an inflammation of one or more tendons, usually at the point of attachment to bone. It is a very common form of soft tissue rheumatism, and you are probably familiar with some of the names given to its specific forms. Examples are *tennis elbow* or *golfer's elbow*, which affect the outer and inner aspects of the elbow respectively. There is also tendinitis of the shoulder, the hip, the knee, the ankle, and so on.[24]

Tendinitis in one specific area is not easily confused with generalized fibrositis, but tendinitis can inexplicably affect many areas simultaneously, causing widespread aching around many joints as well as some degree of muscle stiffness and discomfort. The distinction is that the major areas of tenderness are primarily in the tendons and where they attach to bone rather than in the muscles. Tendinitis, I might add, frequently accompanies fibrositis.

Tendinitis usually responds well to anti-inflammatory medications or local injections of a cortisone derivative. Fibrositis, however, is unresponsive to this type of treatment.

Hypothyroidism

The thyroid gland lies in front of the trachea, just above the sternum, and produces the thyroid hormone. This hormone is one of the major regulators of the body's metabolism. If you have too much, the body "speeds up." The blood pressure and pulse rate rise, weight loss can occur due to the excess burning of calories, you feel abnormally warm, and muscle tremors may

appear. This is called the condition of *hyperthyroid-
ism*. If too little thyroid hormone is produced, the body
"slows down." Blood pressure and pulse rate fall,
weight gain may occur, you feel cold, and your muscles
get stiff and achey. This is the condition of *hypothyroidism* and the one that is important with regard to the
differential diagnosis of fibrositis because of the muscle symptoms it may produce.

Doctors diagnose either too much or too little thyroid hormone in the body in a straightforward manner
by measuring the T3, T4, and TSH in the blood. The
T3 (triiodothyronine) and T4 (tetraiodothyronine) are
two natural forms of the thyroid hormone found in the
bloodstream. The TSH (thyroid stimulating hormone)
is a hormone produced by the pituitary gland in the
brain that causes the thyroid gland to produce T3 and
T4. When the thyroid gland fails to produce T3 and T4,
the pituitary gland responds by producing more TSH
to try to stimulate the thyroid gland to produce more
hormone. So in hypothyroidism, or thyroid failure, the
T3 and T4 are low, and the TSH is high. When all three
tests are normal, thyroid function can be considered
normal. If there is any suspicion of hypothyroidism, the
question can be easily answered by interpreting these
three blood tests. Hypothyroidism can be treated simply and effectively by small daily doses of thyroid hormone.

In the case of fibrositis, however, the thyroid function tests routinely are normal. Despite this, I have
seen many patients with unsuspected fibrositis and
normal or borderline thyroid-function studies who
have been treated with fairly large doses of thyroid
hormone in futile attempts to relieve their muscle
symptoms. The point here is that if you have muscle

aches and pains and normal thyroid function tests, thyroid hormone will not only be of no help to you but could potentially be harmful.

Epstein-Barr Virus Infection

Infectious mononucleosis, caused by infection with the Epstein-Barr (EB) virus, is a well-known, common, acute and self-limiting illness that can last for many weeks. Its symptoms are high fevers, a severe sore throat, swollen lymph nodes, and fatigue. Evidence of past infection with this virus can be found in as many as 90 percent of all adults.

In 1985 two articles appeared in the *Annals of Internal Medicine* suggesting that infection with this ubiquitous virus can also cause chronic fatigue, muscle aches, low-grade fevers, swollen lymph nodes, and an enlarged spleen. Evidence of increased levels in the blood of antibodies to EB virus suggested that this illness was due to a reactivation of the EB virus.[25,26] Interest in the possibility that this virus could cause a chronic condition was heightened when two internists at Incline Village, Nevada, a resort community in the Lake Tahoe area, reported a cluster of patients with similar symptoms seen in 1984 and 1985 to the Centers for Disease Control (CDC).

The illness became known as the Chronic EBV (Epstein-Barr Virus) disease or the chronic mononucleosis syndrome. The media has called it the Raggedy Ann syndrome, since it leaves its victims' muscles floppy, and the Yuppie Flu, since it often strikes the young upwardly mobile professional.

But in 1987 CDC investigators reported they could not clearly demonstrate that the EB virus was the cause of the chronic problem.[27] Indeed, they later officially dropped the name *Chronic Epstein-Barr Virus Syndrome* in favor of *Chronic Fatigue Syndrome*.

The entire controversy had been well publicized in the press. An article in *The New York Times* in July 1987 pointedly summed up the debate over the known significance of the Epstein-Barr virus at the time:[28]

> Medical experts are struggling, with only limited success, to understand a mysterious illness that leaves its victims exhausted for months or years at a time.
>
> But whether patients have fallen into the grip of a new, worsening scourge or have merely succumbed to the latest health hysteria is confounding many medical researchers. And the more experts study patients who have the fatiguing ailment, the less important a culprit the Epstein-Barr virus seems.

The symptoms of this condition bear some similarities to those of fibrositis, although there are actually more dissimilarities. Despite this, many fibrositis sufferers assume that they have the EB virus infection, and many have undergone treatment with potent antiviral agents administered by similarly convinced physicians—with no significant relief at all. But in 1987 an article in *Arthritis and Rheumatism* concluded that there was no convincing evidence to show that EB virus was the primary cause of the chronic fatigue syndrome itself, let alone of fibrositis![29]

The importance of this debate is obvious to the fibrositis sufferer mistaking his or her symptoms for those of an EB virus infection. The approaches to the treatment of these two conditions are as different as night and day. The treatment of fibrositis, as will be explained in Chapter 5, is primarily a natural method, free of the risks that potent medications can pose. The treatment of viral infections, on the other hand, is still in its infancy. A person suffering from an alleged viral infection such as Chronic EB syndrome must either accept the symptoms and let nature take its course or try experimental therapy with a variety of relatively ineffective antiviral drugs that could cause serious adverse reactions.

Yeast Infections

Yeasts are microorganisms that in various forms play many important roles in our lives. For example, some are used to cause fermentation in wine and other alcoholic beverages and some make bread rise. Others, however, can infect humans and produce disease. One of these yeasts, *Candida albicans* (also called *Monilia albicans*), is a normal inhabitant of our gastrointestinal tract but can nevertheless cause infections, known as candidiasis or moniliasis, in susceptible individuals. Some of the more prevalent manifestations are throat infections (also known as thrush) which often affect infants and children, and vaginitis, which often occurs in women who have taken antibiotics.

Within the past few years, some physicians have suggested that hidden infections with *Candida albicans* can release toxins that interfere with the immune system and cause a new array of symptoms and diseases previously considered to be due to other causes. Among these many and diverse symptoms, which have received a lot of attention in the lay press, are chronic muscle aches and pains and chronic fatigue.[30] Proposed therapies have been dietary, plus treatment with antifungal agents such as nystatin and ketoconazole. Nystatin is a relatively benign drug, though it can cause diarrhea, nausea, and vomiting. Ketoconazole, however, is a potentially dangerous drug that can cause permanent liver damage.

What is the evidence that a chronic yeast infection can cause persistent muscle aches, pains, and fatigue similar to those that typify fibrositis? As of now, there are no convincing studies at all to support this claim or to suggest the corollary that anti-Candida therapy will relieve the fibrositis symptoms. In fact, patients who have obvious chronic throat or vaginal yeast infections do not seem to be particularly prone to get muscle aches and pains, and vice versa. The American Academy of Allergy and Immunology has warned that suggestions that yeast infections are the cause of a vast array of purported symptoms are "speculative, unproven and may be dangerous."[31]

PUTTING IT ALL TOGETHER: WHY FIBROSITIS?

Now let's sit back a moment and play armchair detective. We know what generalized fibrositis is—we can identify it with a high degree of certainty. But where does it come from, what are its causes, and what makes it appear? A lot of clues uncovered in this and the previous chapter will help us come up with some answers.

First, we know that generalized fibrositis isn't a new disorder. It has been around for a long time, for hundreds of years, and probably millennia. We know that it isn't a malady peculiar to certain areas or countries of the world—it has a very cosmopolitan distribution. Although it can be found in people of all ages, it has a predilection for striking those in the most productive years of their lives, from ages twenty to fifty. Women are more likely to fall victim than are men, but both can be affected.

Fibrositis does not appear to be an infection from a virus, bacteria, yeast, or any other microorganism. It is not an injury. Unlike many other rheumatic diseases, our immune mechanisms play, if anything, a small part in its genesis.

Unlike many illnesses, it can cause misery for decades and leave no discernible damage to the muscles, or indeed to any organs or tissues. Despite its many symptoms, it does not deform, and it does not kill. It leaves no objective trail—no smoking gun for us to follow.

What, then, might the causes be? Do we have clues and evidence that point toward probable suspects? Indeed we do. Let's look at the company it keeps—the symptoms other than muscular that often, but not invariably, accompany fibrositis. All these disorders—fatigue, headaches, irritable bowel, insomnia, tensions, anxieties and depression—share a common characteristic with fibrositis: In most cases little or nothing can be found physically to account for their appearance. But there is an obvious connection between these symptoms and the symptoms of fibrositis, and we'll explore the relationship in the next chapter.

CHAPTER

3

Fibrositis

BODY, MIND, AND ENVIRONMENT

THE PUZZLE OF FIBROSITIS

In the two preceding chapters we've searched for and uncovered many clues to the causes of and influences on the aches and pains of fibrositis. What we have at this point is a picture puzzle of fibrositis that is still in pieces. We've all worked with puzzles—the picture is on the box, the pieces in the box. We know what the completed picture looks like. Now comes the assembling part, finding out how the pieces interrelate, putting them together and coming up with a clear, recognizable picture.

But our picture of this condition as it occurs in individual people—people with unique bodies and unique life histories—won't be as static as a photograph, a drawing, or a painting, for, as has been suggested, this condition varies greatly among individual sufferers. The muscular discomforts of fibrositis may vary in intensity; they may be persistent or intermittent. The frequently associated symptoms such as headaches, fatigue, irritable bowel, and disturbed sleep may or may not be in the picture. These variations among sufferers of fibrositis are still unaccounted for in our investigations, although we know there must be explanations. These are to be found in the pieces of the puzzle. Let's pick up some of these pieces, examine them, and see how they interrelate.

INSEPARABLE COMPANIONS

Medical science is constantly learning about the causes, meanings, and subjective experiences of illnesses and the meaning and perception of good health. One insight that has come to be keenly appreciated in recent years has particular relevance to the fibrositis syndrome: that body, mind, and environment are truly inseparable companions. The flow of energy that makes up their interaction defines our subjective state of health—our feelings of well- or non–well-being, our comfort or discomfort, our order or disorder, our ease or disease.

Our state of physical health, whether good or bad, has profound effects on our emotional well-being. Conversely, our emotions and state of mind can initiate and alter bodily responses and influence our perception of our bodily sensations. You might easily tolerate a pain of some intensity in your neck and arm, for example, if you think it is due to a strain that will heal and that you might ease it simply by applying heat or taking aspirin. If, however, the source of the same pain is unknown, and becomes chronic, or if you learn that it is due to an uncontrollable spreading cancer, you might find the pain—though of the same intensity—intolerable. Your demand for pain relief would, in that case, most likely increase to the point where stronger medications or even narcotics were needed to bring any comfort. The environment in which you experience the pain can also modify the perceived intensity of it. To be alone in a room or hospital without family or friends might add to your suffering. To be home with strong family ties and support would almost certainly decrease the feeling of misery—and your reliance on medications, no matter what the source of pain.

With respect to this subjective state of health, it's also important to remember that our minds function on two levels, the conscious and subconscious. These two levels aren't always in harmony with each other, and at times they actually conflict sharply. As an example, consider Ted, whose subconscious mind tells him he wants to be dependent and receive attention and protection from someone who cares for him. At the same time, Ted's conscious mind continually asserts his belief in the need to be strong and have full control and mastery of his life. Such internal discord can result in mixed and confusing signals that upset the balance of Ted's body-mind-environment relationship. Somehow, somewhere, in this delicate three-way system, Ted's sense of well-being will be disturbed or distorted in some mental or physical manner by his internal conflict, which in turn will reflect itself in his relationships with those who share his life and environment.

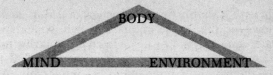

FIGURE 3:1

Figure 3.1 illustrates the complex mind-body-environment relationship. Although the connections shown here have been obvious to Western societies long before modern medicine began to develop, until recently most practicing physicians gave little more than lip service to the practical and therapeutic potentials of focusing on them.

In addition, the rapid and almost incredible advances in our understanding of human physiology, disease, immunology, and genetics in the twentieth century have led physicians into superspecialized and isolated areas of study and practice, with particular emphasis on our bodies and surroundings. In each field, be it for example nutrition, gastroenterology, or cardiology, the knowledge gained, much of it contradictory and open to question, is publicized as fact in immediate press, radio, and television coverage. We are warned at every turn of such dangers as improper diet, infections, and toxins and of the significance and possible serious nature of each and every body sensation we experience. The emphasis is on body and environment, and the mind has been left to react as best it can.

Have all these well-intentioned medical advances made our lives increasingly freer of care, happier, healthier, more content? In the case of the fibrositis syndrome, have they lessened our anxieties about its many uncomfortable and painful but nonfatal and nondeforming symptoms and created an increased sense of well-being and peace of mind?

In a penetrating article in *The New England Journal of Medicine* in 1988 entitled "The Paradox of Health," Dr. Arthur J. Barsky gives us some important insights regarding these questions.[1] He points out that our nation's collective health is excellent and that gains in life expectancy have been "stunning." We have, quite rightly, put an increasing emphasis on the "healthy life-style" accompanied by improved nutrition, weight loss, and physical fitness. Health resorts report increasing numbers of guests, and health fairs are now as popular as boat and car shows.

And yet, Dr. Barksy writes, people report increasingly higher rates of disability, symptoms, and general dissatisfaction with their health. Stated simply, a growing gap exists between our objective health status and our subjective feelings of well-being. Dr. Barksy writes, "According to national polls and community surveys the proportion of Americans who are satisfied with their health and physical condition has fallen from 61 percent in the 1970s to 55 percent in the mid-1980s." One of the major factors appears to be a "progressive decline in our threshold and tolerance of mild disorders and isolated symptoms, along with a greater inclination to view uncomfortable symptoms as pathologic—as signs of disease."

How did all this come about? Here is one of Dr. Barsky's explanations:

> Although our society's fascination with health has had many substantial benefits, it has at the same time eroded the sense of well-being. Paying increased attention to one's body and one's health tends to make one

assess them more negatively, with greater feelings of ill
health. Several investigations have shown that bodily
awareness, self-consciousness, and introspection are
associated with a tendency to amplify somatic symptoms and to report being troubled by more symptoms.
Studies in perceptual and experimental psychology
suggest that, in general, the more aware people are of
their characteristics and attributes, the more negatively
they assess them. This appears to be particularly true
for physical attributes, bodily sensations, and percep-
tions of health.

The paradox, then, that Dr. Barsky has so clearly
pointed out, is that the good intentions of modern medical researchers' emphasis on body and environment
has itself unfortunately created a new environment that
has in certain ways subtly turned against us, tending to
increase our introspective nature and our forebodings
of ill health. Here again we see the mind-body-environment interactions in motion. Information originates
outside our bodies, gets our mind to thinking, and
translates—perhaps not immediately, not with perfect
correspondence—into a subjective feeling of bodily
dis-ease.

As is now commonly known, many ancient Eastern
philosophies and religions have long recognized the
intricate associations of mind, body, and environment
and incorporated them into their philosophical teach-
ings and life-styles to a degree far beyond that of Western societies. For example, for many centuries the
important Hindu philosophy of yoga has taught that
therapeutic effects on body functions can be brought
about by strict adherence to certain of the eight stages
of the yoga process. Among these are *yama* and *ni-
yama*, which stress moral purification and dietary ab-

stentions, particularly through fasting and avoidance of all stimulants and depressants such as tobacco, coffee, and alcohol. Others are *asana,* exercises in posture, *pranayama,* control of breathing, and further stages in which the *yogin* (follower of yoga) mentally isolates and distracts himself from his environment and body sensations through the process of concentrated meditation.

The rigid religious and philosophical aspects of the East are little practiced in this country, since they run counter to our more freewheeling Western ideas. But over the last few decades, an intuitive understanding of many of the therapeutic and healthful effects of Eastern views has invaded our culture. Yoga exercises, for example, have proved to be very effective in relieving internal stress and inducing muscle relaxation and are a welcome and healthy alternative to the massive amounts of tranquilizers and pain medications consumed by our population.

In the last few years, people in the medical establishment have begun to borrow ideas here and there from these traditions and to develop independent but parallel lines of thinking. Meanwhile, research efforts in the laboratories have begun to confirm the mind-body-environment interdependencies so long understood by the Eastern mystics. This turn of events has been crucial to our deeper understanding of fibrositis. Now, with the centrality of these interactions empirically as well as intuitively acknowledged, we have new ways of viewing unexplained muscle pain for what it is, and new avenues of treatment.

Psychophysiologic Relationships

In piecing together our picture of fibrositis, it will be useful to look at some other parts of this three-way pattern, turning to the mind-body interplay, and in particular to the relation of the psyche to body sensations and body functions.

We call the flow of energy from mind to body *psychophysiologic* interaction. The stem *psych-* comes from the Greek word *psyche*, and means "soul" as well as "mind." The term *psychophysiologic* simply designates a physical condition that originates from or is modified by emotional or psychic processes. In fact, it is similar to the term *psychosomatic*. This psychophysiologic reaction can have a variety of beneficial or detrimental effects on our health, feeling of well-being, and performance, depending on our attitudes and states of mind and the consequent physiologic response.

Do we have any real evidence that the psyche can exert a significant influence on both the way our body functions and the way we interpret our bodily sensations? Indeed we do. There is considerable evidence both from empirical studies as on the well-known effects of stress on raising blood pressure, for instance, and the sorts of self-reports doctors hear. Here are a few examples of typical psychophysiological reactions to stressful situations:

Frank, a thirty-four-year-old attorney, always suffered from bouts of anxiety and diarrhea just before he had to take examinations in high school, college, and law school. Now he has the same problems before he has to make a court appearance. No question about it—his nervousness translates into severe gastrointestinal symptoms.

Jan, a twenty-three-year-old secretary, is running a marathon, is ahead all the way, and wins. After she crosses the finish line, a friend points out to her that her leg is cut and bleeding. She recalls slipping and falling but was so excited by the race she felt no pain.

Jim, a forty-two-year-old realtor, gets most of his exercise swimming. When he tries to swim fast, he gets a severe pain in his chest. He is concerned about his heart but delays his visit to his physician, hoping that the pain may just go away. Finally, after constantly ruminating on his condition, developing muscle aches and insomnia, he consults a cardiologist and learns that his heart is fine and the pain is from a slightly irritated cartilage alongside his sternum. His pain now becomes mild and just a nuisance, and his muscle aches and insomnia disappear.

Marilyn, a thirty-nine-year-old professional singer, has problems with headaches, anxiety, a rapid heartbeat, and a flushed face before each concert she gives.

Sarah, a twenty-eight-year-old housewife, gets neck pain and headaches whenever her husband comes home drunk.

Marie, a thirty-five-year-old stockbroker, finds that she relaxes and sleeps much better after having some good laughs watching her favorite comedy show on television.

Sonia, a fifty-year-old secretary, feels weak and numb in her arms and legs whenever her boss criticizes her work.

Jean, a teenager, has a warm and tingly feeling all over whenever her idol appears on television.

Harvey, a high school student, gets sweaty palms when called on in public-speaking class.

Flora, a forty-seven-year-old social worker, finds that her blood pressure is elevated when she is working and normal after two weeks' vacation.

Judy, a twenty-five-year-old teacher, notices that her fingers turn white and cold whenever her class becomes extremely unruly.

Janet, a forty-two-year-old attorney, suffers from severe and chronic muscle pains and fatigue. She is married to a verbally abusive and hostile man whom she finally divorces, at which time her pain and fatigue disappear.

You can probably think of many types of mind-body reactions that you or people you know experience. When the same sensations, symptoms, or physiologic reactions immediately and repeatedly follow transient emotional situations of the same type, the connection between mind and body is often very easy to discern. But when the emotional situation is of long standing and such responses take place chronically—that is, over a long period of time—the connection may not be as obvious and may easily be overlooked.

Mind-body reactions can result in and reinforce a variety of states that we interpret consciously. That is, they are very real to us. When we have positive attitudes, happy emotional experiences, and welcome and challenging stresses, the mind-body reactions can be beneficial, or "asset," effects with respect to our general well-being. When these asset effects of an attitude or experience predominate, we may feel, for example, more relaxed, stronger, more uplifted, or in general exhilarated. There can also be detrimental, or "debit," mind-body effects, which diminish the sense of well-being. And when the debit effects predominate, we may function poorly, suffer from pain, fatigue, physical complaints, or undergo bodily dysfunctions. Such symptoms or sets of symptoms are the *psychophysiologic effects* or even *psychosomatic disorders* described above.

Let's pause here in our account to state a fact not always acknowledged in books on health issues: Many people feel that calling a condition *psychophysiologic* or *psychosomatic* is somehow a criticism or reproach. They feel that a person whose condition is all "in the mind" ought to be "tough" enough or "strong minded" enough to change it. Our new knowledge of mind-body reactions should help to lay to rest this critical point of view once and for all. To begin with, these terms do not mean that the pains and bodily symptoms are "in the mind" or "in the head." They do not mean that the sufferer imagines them or makes them up, possibly for some deep, dark, ulterior motive, or that the person is so weak that he or she has no self-control. And, contrary to many people's interpretation, they are not terms of derision. What they do mean, as you have read, is that the sufferer is experiencing a detrimental, or debit, physiologic response to psychic and emotional stimuli.

Consider the most well-known psychophysiological stress reaction, the fight-or-flight response. This set of body responses consists of a rapid increase in blood pressure, pulse rate, cardiac output, blood sugar, and sweating in the face of danger or a threat of danger. Also, blood is shunted from the intestines to the muscles, and muscle tone increases to prepare the individual to fight or flee. These changes result in part from the release of adrenaline and cortisone hormones from the adrenal glands into the bloodstream, by responses in the sympathetic nervous system, and by hypothalamic stimulation causing release of pituitary hormones. We are now ready for action—ready to respond to a threatening situation.

But what happens when we neither fight nor flee and we still perceive that the threat exists? With the threat present, hormones are still in the bloodstream and the body still in a state of readiness. The energy embodied in the physiological readiness to fight or flee must be burned up rapidly and the tension released; otherwise the body cannot relax. This fight-or-flight response is objective and measurable.

If the stress, which in this case is the subjective experience of danger, persists and finds no outlet, a state of *chronic*—ongoing—stress occurs. The manner in which this state takes its toll on the body depends on our individual constitutions, but take its toll it will, be it fatigue, muscle tension, gastrointestinal disturbances, or disordered sleep. Calling the resultant symptoms "all in the mind" is hardly a productive approach. What is called for instead is acknowledgment and active treatment.

Still, many skeptics demand laboratory or empirical proof that the mind affects the body. Although we have known for years that the stress of sudden and acute danger causes measurable physiologic changes—the fight-or-flight response—experimental evidence was slow in coming to show that chronic or long-lasting stress from any cause or causes could be similarly related to physiologic changes, bodily dysfunctions, disease, and illness. But such evidence is now beginning to appear. An area much investigated is that concerned with coronary artery disease, heart attacks, and hypertension (high blood pressure). In the book *Type A Behavior and Your Heart*, for example, Dr. Meyer Friedman and Dr. Ray Rosenman described the Type A personality, the person who lives in a state of constant self-imposed stress—the person who is constantly planning ahead, who is unable to relax.[2] This behavior has been found to be one of the factors associated with a higher incidence of coronary artery disease and heart attacks. We have a possible physiologic explanation for this phenomenon—sympathetic nervous impulses and excess adrenal secretions appear to be major culprits in the development of coronary heart disease.[3] These same stress-derived responses frequently play important roles in the development of high blood pressure—and, as we shall see, in muscle tension states that result in pain and discomfort.

The picture of interrelationships is growing ever clearer. We are now learning that there are functional links between our nervous, endocrine, and immune systems, and that changes in our immune systems may be influenced by major life stresses. For example, our immune systems protect us from environmentally derived infections—anything from a minor head cold to the fatal acquired immune deficiency syndrome (AIDS). But not everyone is equally susceptible to such attacks, nor are we each consistently susceptible through time. For example, recently bereaved people have an increased susceptibility to illness, and the cause appears to be primarily a psychoimmunologic mechanism—that is, the psyche influencing the immune system.[4]

What is the importance of these new revelations with respect to fibrositis? The answer is that all symptoms, including those generated by fibrositis, can be influenced negatively by debit types of mind-body, or psyche-somatic, reactions. Directing your energies toward clarifying and treating the body or soma alone will be less fruitful than acknowledging the mind-body link in the treatment approach. If you recognize the true nature of the situation, you can use this information to direct your energies toward effecting the *asset* mind-body reactions, by learning and applying methods that reduce emotional or psychic stress contributing to your symptoms.

The Role of Environment

As the mind-body relationship can produce stresses that are internally generated, the environment imposes stresses on the body that are externally generated and often beyond our control. These latter stresses, which have always been important, are assuming even greater roles as our environment changes. Our increasing mobility, changes in family structure, immediate information on changing social conditions and world events, and increasing automation in the workplace are but a few examples of how the world is transforming around us—adding new challenges and stresses to the old ones our minds and bodies must respond to. With regard to fibrositis, these externally generated stresses play very significant roles. In the next chapter I'll describe in more detail some environmental effects on fibrositis, particularly in the workplace. Here, however, I want to continue to focus on subjectivity and its complex relationship to chronic body pain.

THE PERCEPTION OF PAIN

Up to a few decades ago, before the body-mind-environment interaction was given much credence or therapeutic application, most practicing physicians held rather simple and straightforward views of pain and its treatment. "Real" pain was something that really hurt, and was caused by tissue injury, be it trauma, inflammation, or disease. The treatment, in addition to alleviating the source of pain, when possible, was a "painkiller," or analgesic. Aspirin was the first line of defense; if that didn't work, codeine was the next step, followed by tranquilizers and morphine when the pain was severe and intractable. A few anti-inflammatory drugs other than aspirin were available, but they had significant side effects, and physicians were cautious in their use.

Then there was the other kind of pain—pain that could not be explained by some type of injury or disease. Most doctors, as we have seen, considered such pains to be imaginary, in the mind, with no basis in reality, and usually attributed them to hypochondriasis or depression. If they offered treatment, it was usually aspirin or a tranquilizer, reassurance, and a referral to the neurologist, psychiatrist, and, in the case of muscle pains, also to the physical therapist.

Even now, within the context of our new understanding of subjective experience, the pain of generalized or localized fibrositis is often classified in this second drawer—"It's all in your head. Take two aspirin and a nap." One reason that the pain of fibrositis is still shrugged off by many doctors is that it occurs in widely different settings, sometimes in response to an environmental stimulus, sometimes in response, seemingly, to nothing whatsoever external. One can't get a grip on it, can't explain it, can't predict its course. Such pain may follow some type of injury, which in most cases is not severe. I frequently see patients whose fibrositis symptoms appear following a minor automobile accident in which the head and neck are jarred, or after simple strains from tiny traumas—for instance when one reaches around to zip up a dress or lifts a package and "pulls a muscle." Unexpectedly, the symptoms do not improve; the pain persists, worsens, and, in the case of generalized fibrositis, becomes widespread throughout the muscle system. In other instances, however, both generalized and localized fibrositis can appear out of the blue. There is no trauma, tissue damage, disease, or inflammation to account for it. One day, the pain is simply there.

How, in the context of the mind-body-environment model, are we to account for and understand pain occurring in such different ways? Since muscle pain is the primary complaint of these sufferers, we will do well to look to the expanding body of knowledge of pain perception in our effort to account for such pains.

Pain with Injury

First, let's look at one of the widely held theories on how we perceive pain following injury. Experimental studies have shown that various biological products are released at the site of tissue injury. These compounds include such substances as prostaglandins, leukotrienes, histamines, and substance P, which have the ability to stimulate specific pain nerve endings and start a nerve signal on its way as an electrical impulse. This electrical impulse travels along the nerve and enters the dorsal horn of the spinal column. The impulse then travels up the spinal cord and enters a part of the brain called the thalamus. From there, it is referred to the cortex of the brain, at which point it supposedly enters a pain center and becomes consciously perceived as pain.

This train of events—the action—then stimulates a reaction. Compounds called endorphins and enkephalins, which are weak morphinelike painkilling substances, are formed and released in the spinal cord and brain and act to diminish the pain sensations.

Our understanding of this process implies that the severity of the injury should be proportionally related to that of the resultant pain, and this assumption has dominated much of medical thinking. But in the case of fibrositis, there is no such relationship. We need another model to explain the chronic muscle pains of fibrositis associated with injury.

The Gate Theory of Pain

In 1965 Dr. Ronald Melzack and Dr. Patrick D. Wall published a paper that revolutionized our concepts of pain perception.[5] These two researchers realized that the classical theory of pain perception described above was inadequate to explain common observations regarding the presence or absence of pain, for instance, that pain in an injured area could often spread unpredictably to parts of the body where no injury existed, and that extensive injury often did not immediately result in pain. For example, most American soldiers wounded at the Anzio beachhead "entirely denied pain from their extensive wounds or had so little that they did not want any medication to relieve it," presumably because they were overjoyed to leave the battlefield alive. Further, Melzack and Wall noted that no one had even proven the existence of either specific pain nerve fibers devoted to pain transmission alone or a specific pain center in the brain.

The two investigators created a *gate control theory of pain* in an attempt to accommodate these known facts. In part, this new theory is consistent with the classical theory of pain perception. Melzack and Wall proposed that stimuli resulting from tissue injury produce electrical impulses that travel along many nonspecific types of receptor nerve fibers to the spinal cord and then to various parts of the brain, and finally to the frontal cortex where they are consciously perceived as pain. If there is a pain center, they reasoned, then the entire brain must be considered such.

The "gate" part of the theory was this: The two proposed that as impulses travel to the brain, they have to pass a pain gate in an area in the spinal cord known as the substantia gelatinosa, and that this gate can be opened or closed by various factors. Among these factors are variously generated nervous impulses coming from both the skin and the brain itself and the analgesic actions of endorphins and enkephalins.

This theory offers an understandable explanation of how pain can be relieved by temporarily "closing the gate" with techniques that stimulate the skin in or near the area of pain, such as the application of heat or cold, acupuncture, liniments, and electrical stimulation. Also in this context we can understand how emotional reactions generated by positive stimuli—pleasant thoughts or situations; relief and joy at being alive; humor; distraction; pleasing sights, sounds, and smells —can act similarly to alleviate pain. On the other hand, it allows us to understand how negative emotional experiences, such as dwelling on the pain, stress, mental fatigue, anxiety, and depression, can send impulses from the brain to the spinal cord that open the gate and intensify the pain. Finally, the new theory allows us to visualize how positive and negative psychologic responses might affect the production and release of the endorphins and enkephalins.

In an extension of this theory, Dr. Wall gives us some further insights into the perception of pain.[6] In explaining the discrepancy between injury and subsequent intensity of pain, Wall suggests that pain is tied less to the degree of injury than to what he calls a *body sensory experience,* like hunger or thirst. Pain, like other sensations, simply grows worse when unattended. In addition, it is unmeasurable with our present techniques—as are hunger or thirst, for instance— and this, he argues, accounts for our difficulty in assessing even in a ballpark manner what level of pain a person may be experiencing.

Wall describes three postinjury phases—immediate, acute, and chronic. He suggests that the pain perceived in these phases is influenced not simply by electrical currents stimulated by tissue damage and subsequent biochemical events but also by such elements as the environment in which the injury takes place, past and present events, future expectations, emotional states, and distraction; in short, the intricate amalgamation of energies flowing among body, mind, and environment. In relation to chronic pain that persists long after an injury heals, Wall suggests that depression has a central role in the vicious circle that prevents a return to the pain-free state. Wall makes one other observation that is significant to the subject of fibrositis. He states that chronic pain syndromes influenced by the factors named above are often associated with gastrointestinal disturbance, loss of libido, disturbance of sleep patterns, and disturbance of family and social relations—all symptoms and situations remarkably similar to those found in the fibrositis syndrome.

In their pain perception theories, Melzack and Wall articulated what physicians have known and attempted to explain for centuries—that besides direct tissue injury, there is an additional element to pain perception. Psychological and emotional factors can influence not only the severity of the pain of injury but also its very presence. The great baseball player Pete Rose summed up this phenomenon. He had been playing ball with a broken toe and when asked if it hurt, he replied simply, "It doesn't hurt when you're winning." [7]

Chronic Pain Without Injury

Chronic pain that appears in the absence of any known injury or disease has intrigued and fascinated students of pain perception for many years. The fact that such pains are real and not imaginary is not in contention anymore. Given acceptance of the body-mind-environment links, the pains are definitely real. The immediate questions that beg answers, however, are how the pain comes about and what perpetuates it.

Here again we find ample evidence linking psychological and emotional states and the experience of chronic body pain. Stress, tension, anxiety, and depression can all initiate and perpetuate chronic pain in the total absence of any type of injury, tissue damage, or physical disease. Let me offer you just two examples spanning more than a century of material.

Dr. S. Weir Mitchell, a surgeon and neurologist who served in the army during the American Civil War, had some remarkable clinical insights into the interplay between mind and body. Mitchell, a keen observer and prolific writer, had a particular interest in pain, fatigue, and other bodily dysfunctions and their relationship to psychophysiologic reactions. He wrote not only many medical treatises, but also many popular historical and fictional books that often had this mind-body-environment association as the central theme. In 1871 he wrote the fascinating book entitled *Wear and Tear, or Hints for the Overworked,* in which he pointed out the relationship of "cerebral exhaustion" and "brain strain" to such symptoms as muscle pain and fatigue, sleep abnormalities, and body organ dysfunctions—again, the very symptoms associated with the fibrositis syndrome.[8] In one perceptive paragraph, he writes,

Why is it that an excess of physical labor is better borne than a like excess of mental labor? . . . When a man uses his muscles, after a time comes the feeling called fatigue—a sensation always referred to the muscles, and due most probably to the deposit in the tissues of certain substances formed during motor activity. Warned by this weariness, the man takes rest—may indeed be forced to do so; but, unless I am mistaken, he who is intensely using the brain does not feel in the common use of it any sensation referable to the organ itself which warns him that he has taxed it enough. . . . It is only after very long misuse that the brain begins to have means of saying, "I have had enough;" and at this stage the warning comes too often in the shape of some one of the many symptoms which indicate that the organ is already talking with the tongue of disease.

In 1982, 111 years after Mitchell wrote his *Wear and Tear,* Dr. Dietrich Blumer and Dr. Mary Heilbronn reviewed the modern theories of chronic pain, particularly that in the absence of acute tissue damage or disease or ongoing tissue injury.[9] They concluded that we now accept as fact that the brain and psyche are capable of creating chronic pain (in earlier decades before acceptance of the mind-body-environment relationship, the literature was full of hypotheses centered on external causes). They further state that "chronic pain has become recognized as a condition distinct from acute pain and has been referred to 'as a disease in itself rather than a symptom of something else.' "

They view chronic pain as a "specific variant" of a mood state with characteristic clinical and psychological traits. Their studies have shown, once again, that stress, tension, anxieties, depressions, and sleep disorders go hand in hand with chronic pain, and, perhaps even more important, often predate the onset of chronic pain. They observed, as had Mitchell more than one hundred years ago, that many people with these symptoms had constantly driven themselves to work and produce harder and longer than others before the onset of their pain.

THE PSYCHE, STRESS, AND THE FIBROSITIS SYNDROME

As the result of our new understanding of pain, both acute and chronic, we can see fibrositis as a result of a complex interaction involving mind, body, and environment. To find precipitating causes for fibrositis we must look not only to the muscles but to the as yet unmeasurable or poorly measured energies generated by our subconscious and conscious minds and psyches —stress, tension, anxieties, and depression. These forces are at play whether or not injury has occurred. To ignore these energies is to play ostrich and bury our heads in the sand. To recognize them is to come to grips with an enigmatic disorder that has plagued mankind for millennia.

At this point you may well ask why some people get fibrositis and others do not. We all experience periods of psychic unrest. Many people under extreme pressure and emotional turbulence may not get fibrositis, while others, less heavily burdened and strained, may. Many people sustain major physical injuries and do not develop fibrositis, while others with less severe injuries do. How can this be explained? The answer again is simply that people are all different. Let's see how our uniqueness enters the picture.

Target Organs

Our bodies are the products of the complex effects of multiple hereditary units known as genes, which are found at specific points in the chromosomes of our cells. This huge pool of genes transfers genetic material from parent to offspring, providing one factor of an infinite diversity among people.

The study of genetics has shown that our genes not only determine our physical characteristics, such as eye and skin color, but also have profound influences on our emotional, physiologic, metabolic, and immune functions. Further, our genes influence those systems' responses to outside stimuli.

One need not be trained in medical science to know that we each react differently to stress, both that imposed on us from external sources and that we create ourselves. It is also a truism to say that each of us reacts to the same stresses in a variety of ways at different times.

For reasons that are sometimes explainable scientifically and sometimes are not, our makeup determines which organs or functions in our bodies will be affected most—will be the target organs—of stressful situations. Usually the target in each of us is the same. For example, in some people the major target is the cardiovascular system, in others the upper or lower intestines, in others the muscles. Also, most people generally have more than one target organ.

From a treatment standpoint, it is easier and more important to determine which body functions and target organs are being affected by mind-body-environment interactions than to try to decipher the fine details of every nervous pathway or hormone involved in a given symptom. It is a far more direct approach to seek out the root of the problem than to use medications to try to fine tune every complex body reaction. In fact, the latter course often becomes a self-perpetuating and dangerous experience, and a frustrating one for both doctor and patient. Once again, it is easier to control the action than the reaction.

Most people are well aware of what their physiologic response will be to sudden stress. But chronic stress is harder for us to recognize and make predictions about. Its effect on the body is often more subtle than not, making the connection between the stress and the symptoms less obvious and far more difficult to quantify. There is no questioning the fact, however, that for many people the target organs in stressful situations are the muscles, and that the resulting symptoms are muscle tension and pain.

Keep this notion of target organs in reserve as you continue reading; it is a most significant piece in the puzzle of fibrositis. At the same time survey your past with a special eye to possible physical responses to stressful periods. Can you identify your own target organs and reactions?

Somatopsychic Relationships

Just as the psyche can cause somatic or body dysfunctions, physical illness or the misinterpretation of heightened normal body sensations as being a sign of illness can cause psychic unrest. The body-mind relationship can flow two ways.

The fears, anxieties, and concerns generated by symptoms and disease themselves give rise to internal stress and energy that demand release. If this energy buildup continues, the stage is set for the appearance of fibrositis in persons whose muscles are the target organs. If the muscle pains are not identified as fibrositis, both patient and doctor might consider the symptoms signals of a known disease, or consider that yet another disease is also present. In such cases, treatment can be misdirected and wind up being ineffectual and even harmful. Many patients with disorders such as rheumatoid arthritis or lupus erythematosus whose diseases have gone into remission developed the muscle aches and pains of generalized fibrositis because of the concern about their condition. If the situation had remained unrecognized and the symptoms were attributed to the arthritis or lupus, the seemingly logical treatment would be to increase the potent medications used to treat the original disease. This situation indeed occurs often, and with little relief of the complaints. Let patients and doctors take note: Consider the possibility of fibrositis when unexplained muscle aches and pains accompany or follow a physical illness.

Another pain-intensifying situation can arise when the doctor says, "Great news! I can't find a thing wrong with you. Go home and take it easy." From the doctor's point of view, the case is closed. From your own you see a lifetime of chronic pain unalleviated by specific therapy. At that point, according to Dr. Eric J. Cassel in his groundbreaking article, what has been up to now tolerable but troublesome pain may transform itself into unbearable suffering, the distinctive feature being the apparent endlessness of the pain and its adverse effects on your life.[10] It may be, says Dr. Cassel, that your doctor will have little understanding of that suffering and its relation to the deepening of your pain, since medicine has concerned itself so little with the nature and causes of suffering. This lack, he writes, is not a failure of good intentions, since none are more concerned about pain or loss of function than physicians. Instead, it is a failure of knowledge and understanding. It is a major goal of this book to convey this same message to physicians: Chronic pain plus hopelessness equals suffering. Fibrositis—just as sure as heart attack or cancer or muscular dystrophy—can ruin a life, and failure to understand and address that fact on the doctor's part (not to mention that of a spouse or other family member) can transform relentless chronic pain into suffering and despair.

Is There a Fibrositis Personality?

Most physicians concerned with diagnosing and treating generalized fibrositis are familiar with the issue of the fibrositis personality. In fact, there are personality traits that appear to be associated with fibrositis in a higher frequency than in the general population, and this fact lends support to the notion that fibrositis stems directly from stress in certain people whose target organs in times of difficulty are their muscles. Most of the characteristics of a fibrositis personality are ones that, in small doses, we often find desirable, traits that tend to make us good achievers and sensitive to our own and others' feelings. These are characteristics that cause us to impose high standards on our own behavior and that make us reliable, above-average workers.

The following personality traits characterize most fibrositis sufferers:

- perfectionism
- sensitivity to demands made of self and others
- a tendency to repeatedly evaluate the consequences of their actions
- an acute awareness of body sensations
- a tendency toward chronic anxiety and depression

The first four are positive traits that, if exaggerated, cause one to interact with the environment in a stress-creating manner. Such behavior generates a considerable amount of internal energy that must find an outlet.

If these energies and stresses are not sufficiently satisfied or released, the result is increased internal stress, muscle tension, anxiety, fear, uncertainty, and at times variable degrees of supressed anger, hostility, or depression, many of which often predate the symptoms of fibrositis.[11,12]

What kind of a personality do you have? What are

your attitudes toward the daily events of your life? What are your daily thoughts, joys, angers, concerns, hopes, uncertainties, tensions, and stresses? Do you cry often? Are you easily depressed by unfinished tasks, unachieved ambitions, or the insensitivity of those around you? Has your sexual drive decreased? Asking these questions is more than a simple mental exercise for fibrositis sufferers, since the answers will yield you more insight into the cause and treatment avenues of your condition than any combination of laboratory studies and X rays that modern medicine presently has to offer.

I recently asked a fibrositis patient of mine, a very bright and efficient forty-nine-year-old secretary whom I'll call Sandra, if she would jot down a few notes describing her personality. She wrote,

> I'm always in a hurry. On my way to your office I first had to stop and get gas for my car. The attendant was the slowest person in the world. He had two speeds, slow and slower, and I was fidgeting and gritting my teeth. It was all I could do not to get out of the car and push him along. Why I was in such a hurry is beyond me. I am never late to anything. I live by the clock and my husband accuses me of being a human computer. Frankly, I like working under a bit of pressure because I have to work fast and get over with what I'm doing and go on to something else.
>
> My patience limit on a scale of 1 to 10 is 2 1/2. Being a Type A can be fun, but it's exhausting to myself and people around me sometimes. Being this way makes my muscles hurt and my jaws ache because my teeth are clenched together.
>
> I know that sitting all day is detrimental to the body. I try to get up from my desk every half hour or so, or even walk up and down five flights of stairs. I'm starting to do some exercises you prescribed and they

really help. I need to be better disciplined and get into a routine so that I do them every day. Sometimes when I get home I'm so tired and hurting all over that I don't want to do anything but sit down and read the paper. I must admit that when I have exercised and taken one of my long walks with the dogs I really feel well. Even sometimes when I'm really hurting if I just get up and out for a few minutes I feel better.

Even though I'm a Type A person, I don't always express myself, and keep many things bottled up inside. The more negative I feel sometimes, the more I hurt, and I don't want to hurt. I'm aware that only I can control my thoughts if I want to change my circumstances. I'm not totally undisciplined. I want to be well.

Sandra's personality has taken her down a path leading straight to fibrositis. The overriding feature of her life is internally generated stress. In the causation of fibrositis, though, it seems to make little difference whether the stress is generated internally as a result of one's own personality and attitudes, or externally as a result of family, work, or financial pressures. The crux of the matter is not so much the origin or type of stress as how the individual reacts to the stress. Sandra's life is defined by stress, and her fibrositis thrives on it.

VISITING THE MEDICAL SPECIALISTS

Still have trouble believing that all that pain, all that fatigue, all that suffering, all those lost work hours, all those wasted dollars have their major source in commonplace stress, the modern catchall? Consider the experience of the experts. We have specialists in all fields of medicine, and many of the fibrositis sufferers I see have already consulted one or more of them before they reached me: dentists for jaw pains, internists for muscle pain and fatigue, neurologists for muscle pain, headaches, numbness, and tingling, gastroenterologists for irritable bowels, and psychiatrists for the seemingly unrelated constellation of persistent symptoms that puzzle the other physicians.

These specialists usually report that their tests yielded negative results—no physical abnormality could be found. They rule out a serious disease, suspect stress to be the culprit, and refer the patient back to his or her personal physician.

The complaints these patients bring are persistent and troubling. They are just as terrible for the sufferer as they would be if caused by a virus from outer space, and it is reassuring to find out that serious physical disease is not present. The fact is, though, many conditions are known to be caused by stress alone. So if you're still having trouble believing that stress, within the mind-body-environment system, can wreak so much havoc, read this sampling of some common stress-created conditions.

Fatigue

Chronic fatigue and lassitude, unexplained by any apparent physical illness, ranks among the most common complaints that physicians in almost all specialties have to deal with. It is often expressed as a feeling of weakness, feeling all in, losing pep, having no ambition, and the like. Few symptoms generate more referrals to specialists and more laboratory and X-ray investigations. At times a clinical cause is found for fatigue—a hidden infection or malignancy, for example. For the most part, however, the overwhelming cause is rooted not in physical disease but in a mind wearied by stress, anxieties, tensions, and depression.[13]

I recently saw a patient, Nora, a forty-five-year-old married laboratory technician who, for three years, had had overwhelming fatigue, weakness, and muscle pains so intense that she had to give up her job. Her doctor referred her to me after doing a very thorough evaluation and seeking the advice of a host of specialists, the last of whom suggested that Nora suffered from the fibrositis syndrome. Nora brought in the physician's chart detailing the results of his and the specialists' examinations—monthly laboratory studies, blood cultures, investigations for Epstein-Barr virus and a host of other viral studies, multiple X rays, three CAT scans, investigations of her intestines by barium studies and endoscopies, etc.—all of which were perfectly normal. In addition, she brought detailed notes documenting her daily fluctuating symptoms of pain and fatigue over a period of many months. Both Nora and her physicians were convinced that she was suffering from a disease that had eluded their detection.

My examination revealed a healthy woman whose only apparent abnormality was her feeling of pain when her muscles were pressed. The diagnosis fibrositis had appeared in a five-year-old record that I later obtained from one of her previous physicians, a history and diagnosis that Nora's subconscious mind had effectively suppressed. For fully twenty-five years, Nora had been seeking medical help for these same symptoms, which earlier had been present to a lesser degree. The physicians' notes had documented her previous tensions and depressions, along with a life filled with events that had weighed heavily upon her —the tragic loss of a son in a traffic accident, the financial failure of her husband's business, and recurrent concerns about the physical health of her and her family. And now her symptoms had returned with a vengeance, amplifying a vicious and unwanted circle of anxiety, depression, overwhelming fatigue, and muscle pains.

TMJ Syndrome

About ten million Americans, including large numbers of fibrositis sufferers, have chronic pain in the muscles around the jaw, a disorder known as the temporomandibular joint (TMJ) syndrome. Dentists and oral surgeons are usually the specialists most involved in diagnosing and treating this disorder. The cause of 90 percent of the TMJ syndrome is the stress-related localized form of fibrositis known as the myofascial pain syndrome, and the most effective treatment is that related to stress reduction.[14,15]

Headaches

According to the National Headache Foundation, about forty-five million Americans suffer from chronic and recurrent headaches. The most common type of headache is the stress-related muscle tension headache, and some of the most effective treatment programs employ methods related to stress reduction.[16,17]

The Irritable Bowel Syndrome

Millions of people in the United States suffer from the irritable bowel syndrome (IBS). More than 50 percent of outpatient referrals to gastroenterologists are for

management of this problem, which accounts for almost one hundred thousand hospital admissions a year.
IBS is not associated with any organic disease, and researchers have a basic sense that IBS represents some
ill-defined predisposition of the bowel, the target
organ, to abnormal motility. The data suggests that certain definable psychological and sociocultural variables play important roles in those complaining of this
ailment. High on the list of the most effective treatments are those related to the reduction of anxiety and
stress.[18]

Disturbed Sleep

As discussed in Chapter 2, disturbed sleep is often a
prominent part of the fibrositis syndrome, one that may
even result in a referral to a sleep disorder center.
Whether it plays a primary causative role or is a sec-
ondary reaction to chronic pain is not clear. What is
clear, however, is that stress and anxiety can be a major
factor in producing insomnia, even in individuals who
are otherwise perfectly healthy.[19] In addition, lack of
proper sleep can alone cause muscle aches and pains.
Once again we see the role of the psyche in adding fuel
to the fire of the fibrositis syndrome.

PUTTING THE PUZZLE TOGETHER

At this point, we can stop and ask ourselves whether
we have enough pieces of the puzzle to put together a
clear picture of generalized fibrositis. The answer is
yes. Some pieces are missing, but with those we have,

the condition has become easily recognizable—clear enough to allow us to diagnose and treat it.

When the pieces we have are in place, we see a picture of muscle aches, pains, and associated symptoms created by a complex interaction of mind, body, and environment, with the muscles themselves showing no apparent abnormality. We see a picture of a person whose major target areas under stress are and usually have been the muscles. In this susceptible individual, externally or internally generated stress stimulates an internal psychic energy force that

- causes muscle tension
- creates a sense of fatigue
- decreases the feeling of well-being
- lowers the mind's tolerance to pain
- amplifies and distorts the mind's perception of body sensations
- interferes with normal sleep patterns
- has the ability to cause a host of other psychophysiologic responses
- can become despair when the sufferer sees no end in sight

It is the personality of the sufferer that is the central factor behind these psychic energy forces. Here the term personality encompasses the person's reactions to life's events—his or her hopes, fears, anxieties, and concerns. In this particular human being, interactions with the environment that produce negative emotions and feelings can quite directly trigger muscle pains.

What pieces are missing from this picture and how serious a problem does that pose? The missing pieces are our knowledge of most of the subtle and complex biochemical and bioelectric links that transmit the energies back and forth from mind to body. This is knowledge that no doubt we will eventually gain at least in part, but our current lack of it neither obscures our picture of fibrositis nor seriously diminishes our ability to deal with its causes.

CHAPTER

4

Localized Fibrositis and Strain Injuries

ANOTHER PUZZLE? THE FIBROSITIS LINK

Over the past few decades, while physicians and researchers were trying to put the puzzle of generalized fibrositis together, another picture puzzle was forming. Although some were known for centuries, various and confusingly named localized muscle pain syndromes, illnesses affecting specific parts of the body, were and still are being reported from all over the world with a rapidly increasing frequency.

The two forms of fibrositis, generalized and localized, are basically similar in origin and treatment. I have separated them in this book for three reasons. The first is that localized fibrositis is emerging as an important work-related issue, the workplace defined as either an industrial or nonindustrial setting (such as the home). The second is that in localized fibrositis physical events, such as muscle stresses, preceding the onset of pains are often more clearly defined. The third is that the varied pictures of localized fibrositis can give us further insights into the generalized form.

Again, practicing physicians have found these pains difficult to understand, since despite any history of alleged muscle stress or injury, objective physical findings have rarely been found to account for them. As with generalized chronic muscle pain, treatment has usually been based on the assumption that these problems are caused by as yet unidentified localized or specific physical problems, such as muscle irritation or damage, or associated inflammation of tendons (tendinitis), and has been frustratingly ineffective. Perhaps because of the perplexing nature of the illnesses, most general physicians have relinquished the investigation and care of patients to rheumatologists, neurologists, orthopedic surgeons, physiatrists, industrially based physicians, and dentists—those specialists whose areas of expertise seemingly were strongest in uncovering and treating the alleged cause.

The major areas of research into these localized muscle pains have been reminiscent of paths taken earlier in the investigation of generalized fibrositis—concerns about trigger points, arthritis, muscle strain, nerve damage, evaluation of X rays and laboratory studies, and the like—with only very minimal consideration of the total mind-body-environment relationships. This primarily body-oriented approach, unfortunately, has not clarified much of the mystery. These illnesses have proven neither easy to understand nor simple to treat on a purely physical basis.

But we can take a broader and far more effective approach simply by applying the lessons we have learned about generalized muscle pains. These localized muscle pains are best understood in the same context of body-mind-environment relationships. In fact, it is useful and accurate to think of them as another part of the fibrositis picture puzzle—a localized form of fibrositis.

Consider three case histories:

For Carol, a thirty-four-year-old housewife, neck and shoulder blade pain began on her wedding day, immediately after she reached back to zip up her dress. The pain was still there when she first came to my office, a few days after her tenth wedding anniversary! Despite many previous examinations and X rays, no physical cause had been found to account for her pain—a pain so severe that Carol was unable to spend much time at certain household chores such as cooking, cleaning, and doing the laundry, which eventually created some anger in her husband. The pain was present at night to the point where she had to position herself with her arms spread out, requiring her husband to sleep in another bed, wreaking havoc with their sex life. The pain seemed the catalyst that allowed the antagonistic interactions between the two to proceed. The case, however, proved to be otherwise. Marital counseling finally brought out the fact that the antagonisms had begun during their engagement. As counseling continued, they fortunately were able to come to a mutual understanding, and her pain faded into the background to the point where it no longer needed any consideration of treatment.

Janice, an intelligent, extremely perfectionist, and highly motivated thirty-eight-year-old woman, had

had minor pains in her neck, shoulder, and right arm since her teens. In her early twenties, she went to work for a newspaper and soon became the head of the newsroom, only to lose her job at age thirty-five when the company was sold. She then married and, at the insistence of her husband, gave up her career and assumed the duties of a housewife. At this point, the localized pain greatly increased and spread through-out her body in a typical generalized fibrositis pat-tern. She sought the help of twelve specialists, only to have her pains increase to the point where she had trouble sleeping and had to prop her arms on pillows to get comfortable. Her husband had no room left in the bed and began sleeping in another room. What was the problem? She told me what previously she had told only a few close friends. Although she loved her husband, she felt that their present acquaintances had narrow interests, and she longed for the intellec-tual stimulation she previously had. She saw her ca-reer slipping by, and felt like she was in a trap. In addition, her husband desperately wanted a child, and she did not. She saw motherhood as the final nail in the coffin that would bury her career aspirations forever. She admitted to being depressed and having crying spells. The situation here did not call for more blood tests, X rays, and examinations. They had been done many times over and had always proven to be normal. Until Janice and her husband were able to reach a mutual agreement on their goals, her pains were destined to continue, along with her anxieties and depression.

Mel, a thirty-seven-year-old accountant, had the onset of severe and persistent neck, right shoulder, and arm pain and mild depression at the age of thirty during a very stressful period in his life. He had joined a small but growing company and spent many hours at a computer daily without taking a break. He was treated with stress-reduction techniques, postural and stretching exercises, and, for a short period of time, an antidepressant medication named amitriptyline, which completely alleviated his symptoms. When Mel was thirty-four, the company went through a period of financial crisis. He was again working long hours, when he had the onset of jaw pain typical of the TMJ syndrome. His dentist diagnosed malocclusion, and over the course of two years Mel had four teeth removed, and treatment with braces and a splint—all to no avail. He sought another opinion from an oral surgeon, who said he still had malocclusion and proposed doing surgery on his jaw in order to improve his bite. At this point, he saw me to ask my opinion. I suggested that before any surgery, he should consider his pain as a reaction to the stressful situation he was in and restart the treatment program that he had undertaken for his neck and arm pains. Within three months his TMJ syndrome disappeared.

Here, in these three people, we see stress as the primary cause of localized fibrositis. Carol's recovery hinged upon the realization that marital stresses were major contributing factors to her symptoms. In Janice's case her continuing anxieties and depression were not only at the core of her localized fibrositis but had led to her generalized fibrositis. In both situations, the workplace was the home.

In Mel's situation, it is almost certain that an earlier appreciation of his muscle stresses caused by typing on a computer keyboard and of his job tensions would have led to a more rapid resolution of his problems and saved him much pain and money.

There are, once again, many paths to localized fibrositis. In some instances, the body predominates and the major cause is repeated physical muscle stress. In others internal stress generated by the mind is the culprit. In still other instances, the stress may come primarily from the environment. Finally, all three factors may assume major roles.

The point is that if you suffer from localized fibrositis that is not easily and successfully treated, you have to think beyond simple muscle strain and injury and ask yourself how much internal stress or your reaction to the situation around you may be contributing to your symptoms. If your physician, or dentist in the case of the TMJ syndrome, does not consider and ask about these same questions, critical pieces of the puzzle will be missing, and your treatment will be incomplete, misdirected, or possibly harmful.

Defining Localized Fibrositis

To sum up this condition, localized fibrositis is chronic persistent pain and tenderness affecting muscles in a localized area of the body, such as the neck on one side, a shoulder or shoulder blade, one of the arms, forearms, or legs, the back or the chest, the low back, or various combinations of these. Any of the muscles that support the spine and are responsible for posture can be involved. The most common form is characterized by a diffuse aching and stiffness from the lower portion of the neck, through the shoulder, upper chest and into the arm and hand. As with generalized fibrositis, the symptoms may be associated with sleep disturbances, stress, irritable bowels, and headache, and they are usually aggravated by use of the affected area, weather changes, and emotional stress.

Like generalized fibrositis, the localized forms also can be accompanied by subjective feelings of numbness, tingling, burning, weakness, and heat or cold in the affected area. The distribution of the pain is diffuse and usually corresponds to muscle groups rather than to any specific joint or nerve distribution.

Just about anyone can fall victim to localized fibrositis. It can occur from childhood to the limits of old age, but people in the most active and productive years of their lives—from their twenties to their fifties—are most susceptible. As with generalized fibrositis, the localized form strikes women more than men.

Searching for Causes

The syndrome of localized fibrositis has been known for many years, albeit under a host of names. One name used in the United States is the myofascial pain syndrome, a term used to differentiate localized from generalized chronic muscle pain and one that was popularized during the 1950s.[1,2,3] Dr. Janet G. Travell and Dr. J. J. Bonica had put forth the idea that these pains were due to muscle trigger points, often accompanied by some abnormality felt in the muscle consistency (the fibrositic nodule), which I discussed in the previous chapters, and thought that the chronic pain resulted from a pain—or muscle spasm circle. A physical origin for the trigger points has never been demonstrated, however.

When it has been described by physicians from this and other countries, however, the concept of trigger points and fibrositic nodules has not always been accepted, and often is not mentioned at all. Outside the United States it is usually given a name other than the myofascial pain syndrome. Despite the easy exchange of medical literature among countries, nationalism or some other overriding factors inspire physicians with the notion that what they are seeing is due to some entirely new and as yet undetermined cause, is peculiar to their own region or country, and is deserving of a special name. However, whatever the reasons may be, the localized fibrositis conditions are remarkably similar from one country to the other, and it is reasonable to assume that the causes are similar.

Most physicians around the world are in agreement on one point—a history of muscle stress, be it from constant contraction, repetitive muscle movement and strain, or chronic muscle tension from psychic stress, often precedes the onset of chronic neck and arm pains. Examples of activities often associated with the condition are playing musical instruments, using typewriters or computer keyboards, assembly-line work, knitting, or even watching television for long hours. Also, it is a universal observation that even a single muscle strain or injury, whether insignificant or serious, initially, may be followed by localized, then generalized fibrositis.

Again, the muscles that support and move the jaw joints are common targets for localized fibrositis. This condition, which is popularly known as the *temporomandibular joint (TMJ) syndrome* or the myofascial pain syndrome (MPS), or myofascial pain dysfunction (MPD) of the jaw, causes pain in the jaw joint and the surrounding jaw muscles when the person is chewing or opening the mouth. Headaches and earaches may accompany this syndrome. Arthritis of the jaw joints or malocclusion—improper meeting of the upper and lower teeth—had formerly been thought to be the most important causes. It is now well-known that increased stress can increase tension in the jaw muscles, and there is considerable evidence to show that psychological factors such as anxiety, depression, and stress are most important in the causation, progression, and treatment of the TMJ syndrome. Also, there is a growing recognition that much major dental reconstructive therapy, resulting in irreversible mechanical changes in the jaw, is unnecessary and poses a potential danger to some patients.[4]

Other painful and often disabling musculoskeletal disorders have been described under at least 150 names, most incorporating terms related to the involved area or the assumed origin, and any experienced rheumatologist could add more. One condition, for example, has been called the *ponderous purse syndrome*, and is brought on by carrying a heavy purse with a shoulder strap. The weight causes constant contraction of the shoulder and neck muscles, followed by painful muscular contractions. Another condition has been named *back-pocket sciatica*—this is a pain in the buttock and leg brought on by constant pressure and irritation of muscles from a thick wallet carried in the back pocket.[5] Another that I frequently encounter is chronic neck and arm pain precipitated by long periods of holding a telephone receiver to the ear with the shoulder.

A recently recognized feature of localized fibrositis is that there have been epidemics of it in industrial settings where the same type of work has been going on for decades or even hundreds of years with no apparent major problem. Here again we are learning how both psychogenic energies and the environment in which the pains appear can be major determinants behind the causes and perpetuation of these muscle pains.

The Tower of Babel Revisited

I'm going to return to the idea of the Tower of Babel to make a statement that may surprise you: If you have fibrositis, either local or generalized, the name your doctor uses in the diagnosis may well affect the course of your condition, the treatment you receive, and any reimbursement you may receive in the form of insurance, disability payments, legal settlements, and the like. The labeling of localized fibrositis symptoms is more than just an intellectual exercise in nomenclature. It reflects your physician's concept of the origin of your pain. Let's consider a few examples.

If a physician or insurance examiner does not believe that you are really in pain, you might be given the diagnosis of *malingerer* or another euphemistic term. There are surely cases where a person pretends to be ill or incapacitated to avoid work, to collect disability payments, to defraud an employer, or to collect a large settlement in a legal suit. If you are not one of these pretenders, you might react to such a label with anger, hostility, depression, or even doubts about your own sanity. These feelings will not help you get rid of the pains. On the contrary, they will undoubtedly intensify them.

Another possible diagnosis for your chronic local muscle pains is *psychogenic rheumatism*. This term implies that the pains are all in your mind, not real, and that your muscles really don't hurt. The odds are that with this diagnosis you will be shunted off to a psychiatrist or psychologist who, despite his or her expertise in probing the psyche, will probably know little or nothing about fibrositis. Your reaction might be, "They 'sort of' believe me, but not quite. I *know* that I have pain and it isn't in my mind." Again, you will probably not get much better, and probably will get worse. Remember that, in Dr. Cassel's definition, pain plus hopelessness regarding relief equals suffering.

A name that includes such terms as *injury, trauma, overuse* or *strain* implies that there has been some form of tissue damage and focuses attention and treatment primarily on the results of the supposed injury, which may or may not have occurred. This sort of diagnosis can have two important consequences: First, in the mind of your physician, the relationship of mind-body-environment to cause and treatment will be relegated to the background, if it is considered at all. Second, any insurance company or worker's compensation board involved will view compensation differently on the basis of whether or not an injury was involved. More time and energy will probably be spent trying to settle this issue than in treating your aches and pains. The effect of the subsequent turmoil on your psyche and pains will not be a good one.

The ultimate goal of the patient-physician relationship should be the relief of the patient's suffering. There are some labels that reflect this objective—namely *localized fibrositis, regional pain syndrome,* or *myofascial pain syndrome,* and are preferable. These terms do not imply a single specific cause. Rather, they allow patient, physician, and anyone else involved to explore freely all the contributing causes in order to come up with the most effective treatments.

FIBROSITIS, STRAIN INJURIES, AND THE WORKPLACE

We now know with certainty that the aches and pains of localized fibrositis are molded and influenced by varying social, occupational, environmental, and economic as well as psychological, emotional, and physiologic factors. Nowhere is this more apparent than in the industrial workplace, and most often even the best physicians are poorly equipped to deal with the situation. In fact, such factors are often related to the very structure of society in industrialized countries. In a very real sense, an attack of fibrositis could be viewed as a result of treating a finey tuned organism—the human being—as merely another tool in the factory.

I cannot pretend to have solutions to the social, industrial, and economic problems that plague our planet, but I can point out the links between these problems and the causes and prevention of fibrositis. As we look at fibrositis in the workplace, it will become clear why this illness rarely responds adequately to a treatment program consisting only of medications and massage.

Work Disability

The disorders of the musculoskeletal system—both arthritis and the soft tissue rheumatic disorders—are second only to respiratory illnesses as a cause of work disability. Among the chronic diseases, they rank second in the frequency of physician visits.[6]

Most of the studies done on the social and economic effects of the musculoskeletal diseases have been concerned with osteoarthritis (degenerative arthritis), rheumatoid arthritis, and low back problems. Only recently has fibrositis drawn similar attention. In fact, it has only been very recently that the medical profession, industry, insurance companies, and governmental agencies have begun to look, or even glance, at these socioeconomic-fibrositis links. The impetus for this new interest is due in part to the resurgence of medical interest in fibrositis, but mainly to the tremendous rise in the cost of medical care, health insurance, and work disability.

Recognition of the full medical, social, and economic implications of fibrositis and the other musculoskeletal illnesses has been slow in coming in the industrialized Western world. But although we still have much to learn, one area of investigation in particular, that of work disability, has helped us to understand fibrositis and its global and individual consequences. To help you understand these supposedly enigmatic muscle pains, I'm going to take you on a journey to the workplaces of Japan, then Australia, and back to the United States.

The Japanese Experience

Japan, in 1971, was the first country in which a major concerted and directed governmental effort was made to define the problem of localized fibrositis in the workplace.

In Japan, the 1950s and 1960s had seen a rapid development in the mechanization and automation of many industries, and this led to a distinct increase in the performance of simple and repetitive tasks by workers. This development, in turn, resulted in an increasing type of occupational health hazard, particularly among young female workers, that the Japanese call the *occupational cervicobrachial disorder* (the same term is also used in Scandinavia).[7] If we climb back onto the Tower of Babel for a moment, we find that this disorder is identical to that known in the United States as the myofascial pain syndrome, a form of localized fibrositis described in Chapter 3.

In 1971 the Japan Association of Industrial Health organized a committee to deal with the increasing number of workers affected by this painful muscular problem. The committee found that most of the afflicted people worked in certain industries, particularly those using keypunch machines such as computers, telephone terminals, and supermarket cash registers. The investigators were able to isolate several major factors that they felt were definitely related to the cause of this disorder:

- An excessive muscle load on the upper extremities from repetitive activities, or from static muscle work such as keeping the arms raised in one position for a long time, causing chronic fatigue of the muscles.
- Uncomfortable working postures.
- Mental stress and severe tension.
- Environmental factors such as noise, uncomfortable temperatures, and poor lighting.
- Poor working conditions due to personnel management.

The importance of the committee's report lies not so much in the description of symptoms, which had been well-known for years, but in the recognition that the *type* of muscular activity, the *place* in which it occurs, and the *mental stresses* involved are important causative factors. The Japanese found that the occupational cervicobrachial disorder was indeed a complex result of mind-body-environment relationships. Their investigations have continued, and have served as the model for studies in many other industrialized countries.

The Australian Experience

Australia has become the major center for the discussion of the causes and treatment of localized fibrositis. There, discussion of the issues of these muscle pains has reached an almost feverish pitch. And there the general public, unions, disability boards, and attorneys have all joined the fray. In Australia physicians have become vehemently polarized in their views and debate the subject with a frankness unmatched anywhere else in the world. Let's take an armchair trip to Australia and see for ourselves.

The Problem

In 1984, it became apparent to the Australians that they had "a major unchecked source of disability in industry and commerce" on their hands—a rapidly increasing number of cases of localized muscle pain paralleled by a rapidly rising rate of compensation claims. In New South Wales, for example, the rate of compensation claims rose 220 percent over the period from 1970 to 1979.[8] Many employers were shocked to find that work-

ers' compensation premiums had risen between 100 to
300 percent in a two-year period in the early 1980s,
representing an expenditure for many employers that
was second only to wages. In 1981 the workers' compensation insurers sustained a loss of 243 million Australian dollars, and many found that these muscle
conditions were second in cost after back injuries.[9] By
1984 the situation warranted the title of The New Industrial Epidemic. The author of one article pointed
out that the reported incidence represented only those
cases reported, and that many go unreported for many
reasons.[10]

As a result of the growing concern, the National
Occupational Health and Safety Commission
(NOHSC) established the Repetition Strain Injury
(RSI) Committee to investigate the problem in October
1984. Their superb final report in 1986, "Repetition
Strain Injury: A Report and Model Code of Practice,"
provided a detailed insight into this illness.[11]

Localized fibrositis is known by a number of names
in Australia, but the occupationally induced form is
most commonly referred to as a *repetitive* or *repetition
strain injury* or simply *RSI*. The term *RSI* actually encompasses various other soft tissue rheumatic illnesses
as well. Among such disorders are pains due to inflammation of the tendons in the hands and wrists and the
carpal tunnel syndrome. This latter ailment is characterized by various combinations of pain, tingling, and
numbness in the palm of the hand, often radiating up
the forearm, and is due to pressure on the median
nerve in the wrist area. Localized fibrositis, consisting
of less well-defined pains, actually comprises the great
majority of cases, however. I might add that many Australian physicians, noting that in most cases no observ-

able injury precedes this disorder, prefer the terms *repetitive strain syndrome—RSS—*or *occupational overuse syndrome* rather than *injury,* adding still more terms. The general public and much of the press in Australia, picking up on the initial reports that some people suffering from this disorder had inflammation of tendons and their sheaths (tendinitis or tenosynovitis), have adopted the abbreviated name of *teno.*

The Findings

The NOHSC report was extensive, and I will simply offer some comments on its highlights:

- RSI is not an Australian Disease, as many Australians had initially thought, nor is it new. It is a well-known and major problem under investigation in such industrialized countries as Japan, Sweden, Norway, Switzerland, Germany, and the United States, and the recognition of RSI-type symptoms goes back several hundred years.
- Females are affected in a higher incidence than males, as noted in reports from Australia and other countries as well.
- The report confirmed the Japanese findings that various types of muscle stresses and strains can lead to RSI. These muscle stresses can occur in almost any work situation, from light office clerical jobs to strenuous or prolonged repetitive labor.
- RSI is not a single type of illness but rather is a collection of muscle and tendon problems. A small percentage of these illnesses are clearly associated with inflammation, such as tendi-

nitis and bursitis, and are well defined with clear-cut treatment options. In the majority of cases, however, no evidence of tissue injury or inflammation can be found.

- Despite the known absence of injury in most cases, the committee (surprisingly, I believe) agreed to adopt the name *repetition strain injury,* with *occupational overuse syndrome* running a close second.

- The report repeatedly referred to the importance of psychological and psychosocial factors in the causation, progression, and treatment of RSI. Among these factors were the effect of chronic pain and disability, psychological factors in the workplace and personal social life, and the adversarial nature of the medicolegal process of workers' compensation. It also noted that "attitudes to sufferers of RSI influenced their perception of themselves and, as a result, the outcome of the conditions." Those attitudes would include those of family, physician, and employer.

- Suggested preventive strategies were based on the fact that a strictly physical approach to these localized muscle illnesses had proven to be inadequate, and that *ergonomic* principles should be embraced. These principles form the basis of the relatively new international specialty of ergonomics. (Ergonomics is defined as the study of human performance at work. Its aim is to promote the well-being, safety, and efficiency of the worker by study of his or her capabilities and limitations in relation to the work system, machine, or task, and in relation to the physical, psychological, and social environment in which he or she works. It is concerned with the interaction of the three main elements in any work situation —person, machine, and environment.

The Controversy

The NOHSC report acknowledged the controversy that was beginning to swirl around RSI in the early 1980s, and pointed out that marked differences of opinion on this condition were reflected in letters to the editors of Australian national newspapers, cartoons, feature articles, and professional journals. Significantly, the report noted, "This polarization of attitudes has influenced the outcomes of RSI for workers, employers, health professionals, insurance companies, legal practitioners, unions, co-workers, and family and friends."

An important thrust of the report, then, was that these contrasting attitudes were not just matters for academic discussion but were to be considered matters of practical importance in the diagnosis, treatment, and outcome of localized fibrositis.

Perhaps the course of your own fibrositis may have been influenced in one way or the other by the attitude of your physician or others, and it may be enlightening for you to peer in on one scene of a microcosmic controversy that took place in 1986, the same year that the final NOHSC report was released. This controversy involved the musicians of Australia and RSI.

In 1986 an article by Dr. Hunter J. H. Fry, a plastic surgeon and accomplished musician, appeared in *The Medical Journal of Australia*, reporting a series of 379 musicians with RSI of the neck and upper limbs.[12] The condition, he found, could affect a player of any type of musical instrument but was particularly prevalent among string, woodwind, and keyboard players. More than 50 percent of members of orchestras and up to 20 percent of students suffered the condition of localized fibrositis. His preferred method of treatment was "radical rest of the tender structures by the total avoidance of pain inducing activities." The symptoms could be prevented, he stated, by better positioning and shorter periods of playing. Dr. Fry felt that very few musicians were prepared to talk about their condition because they feared they would lose their place in what was a very competitive industry.[13] Within a few months, it became apparent that Dr. Fry's report had touched some very raw nerves and evoked strong reactions as exemplified by letters to the editor in *The Medical Journal of Australia.*[14]

On the one hand, there were unsympathetic physicians who claimed that the stated number of legitimate RSI cases among musicians was blown totally out of proportion, if indeed most existed at all. Their criticisms were based mainly on what was a purely body-oriented approach—that is, the inability of anyone to find clear-cut evidence of either injury or physical abnormalities. Also, they felt that this epidemic seemed peculiarly confined to Australia (which, indeed, it was not). Dr. Fry's report was criticized unmercifully, albeit colorfully, as a "farrago of heuristic assertions and unsophisticated hypotheses," a "sham of iconoclasm," an "intermezzo of his *opera buffa,* but . . . the libretto has an unconvincing plot," and that it illustrated a "form of mass hysteria." It was suggested that large court settlements and liberal workers' compensation laws in Australia were leading to or reinforcing what was referred to as an occupational neurosis. One physician went so far as to write, "I think we are doing these patients a disservice by attempting to treat them. It tends to make patients honestly believe there is something seriously wrong with them and tends to make them think that they will never return to the workforce. . . . We have to call a halt to this nonsense somewhere." This is the same old song again—"It's all in your mind. Take an aspirin and be thankful there's nothing *really* wrong with you."

On the other hand, Dr. Fry was applauded by both musicians and physicians alike for a "significant advance in our knowledge of the occupational overuse syndrome." Some letters denounced, often from personal experience, many physicians' ignorance of this illness and pointed out that the pains were very real to those who suffered from them. It was also claimed that physician neglect and rejection by employers and insurance companies caused many patients to become demoralized, dejected, and reclusive. One physician writer said that for many patients dejection is quite appropriate, considering the confusing messages that they received from various physicians and authorities regarding treatment. He wrote, "in fact, many patients find themselves wearing splints for one doctor, being told to take those splints off by another, being recommended back to work by one authority, and being told under no circumstances to use electronic keys by another authority." He wisely recommended that the medical profession put its house in order by further studying the cause and treatment of this obviously real condition and ceasing to treat these patients as if they were neurotic. We need only remember Dr. Cassel's description of the role of hopelessness in exacerbating pain to applaud this recommendation.

In 1986 Dr. Geoffrey O. Littlejohn offered a factual and realistic summary of the Australian experience— an experience that has remarkable parallels in other industrialized countries.[15] He pointed out that the repetitive strain syndrome is a localized and exaggerated form of the fibrositis syndrome and is very often precipitated by specific "emotion charged life events." The causative factors, he notes, relate to a complex interplay of various psychosocial, medicolegal and medical management factors. He further wrote that the Australian epidemic of localized fibrositis can teach us much that pertains to generalized fibrositis, and that to comprehend both we must have an understanding of the relationship of stress to human behavior.

The American Experience

Neck and arm pains have been recognized as major health problems in industry in the United States since the mid-1960s, although physicians, employers, and insurance companies have shown little interest in these conditions until recently and then only minimally. There have been some attempts in the United States to define the magnitude of this medical challenge, but they have been hampered by our inability to clearly define the work situations in which they occur, to find any consistent physical abnormalities to account for the pains, or even to come up with an acceptable name for the group of illnesses.

In 1977 Dr. Nortin M. Hadler coined the term *industrial rheumatology* in an attempt to apply the expertise of rheumatologists to arm and neck problems and to bring some order out of chaos.[16] Recognizing that these disorders cut across all types of pursuits, both industrial and nonindustrial, Hadler wrote that "we are not studying disease of industry but of people." [17] In 1984 the United States National Institute for Occupational Safety and Health (NIOSH) drafted a five-year program and targeted money to investigate the extent and causes of what it called *cumulative trauma disorders (CTD)*—yet another term to add to the confusion.[18] But this effort resulted in only a handful of medical articles and then only in specialty journals rarely read by most practicing physicians.

Other forces have been at work, however. Following the Australian lead, many organizations in the United States have sponsored conferences on repetitive strain injuries. Also, a bimonthly news report, *VDT News,* now publishes information on health and safety issues associated with computer use.[19]

A particularly important turn of events in this regard occurred on June 14, 1988, in Suffolk County, New York.[20] The county passed a law, the first in the nation, regulating working conditions for video display terminal operators. The regulations included requirements for eye examinations, work breaks, and workstation standards, for example, the type of chairs and tables to be used and proper lighting. Local industries and businesses are still fighting this law bitterly, claiming that the expenses of such regulations would put them at an economic disadvantage with businesses in surrounding counties. This case, I believe, portends an avalanche of similar battles in which health measures are pitted against economic factors.

Medicolegal Issues

Workers' compensation statutes, and various forms of disability insurance in the United States and other countries, have protected many individuals against poverty following disabling injuries or illnesses. These well-meaning, very real, and effective programs have, nevertheless, given rise to other problems. When the symptoms of localized fibrositis are deemed "job-related" by the diagnosing physician, all too often an economic battle results involving the employer, the afflicted employee, and the insurance companies. All the stresses and strains of financial warfare ensue. Such disability cases can hang in limbo or linger in the courts for many years, during which time there is little incentive for the employer either to admit responsibility for unhealthy working conditions or to change them. Nor is there cause for the fibrositis patient to improve. At this point, the physician, as I can attest from personal experience, soon realizes how limited is his or her role in the patient's total recovery. The eventual losers in these very costly confrontations are the patients and their families, industry, and the public at large.

In the United States in 1985, Dr. Nortin M. Hadler offered these comments on localized fibrositis and its relationship to the workplace and disability. He wrote,

> The illness of work incapacity is experienced in a complex sociopolitical climate so laden with anger, invective, vested interest, bureaucracy, costs, and profit taking that it serves the afflicted unevenly and inefficiently. The system can confound the best efforts of a well-intentioned and knowledgeable primary physician.[17]

In other words, the well-intentioned efforts at reaching a resolution to such battles couldn't be better designed to intensify the pains in question. Compensation battles breed stress, and the stress plus the target muscle groups equal an intensification of the symptoms of fibrositis.

Let me offer you a not too unusual example of this situation.

A Case Report

When I first saw Margie, a laboratory technician, in my office, she was twenty-seven years old and had been receiving state disability payments. About two and a half years before she came to me, she had noted some aches and pains in her neck and right arm, which she attributed to looking through a microscope at work for many long hours a day. Her personal physician, whom she had seen for this problem on two occasions, had diagnosed her problem as a muscle strain, and treated her with aspirin, a muscle relaxant, and increased rest, with very little improvement.

Two years before I first saw her, she was driving a car on company business and was involved in a minor automobile accident in which a fender was slightly dented. She had not suffered any physical injury at the time of the accident but was quite upset over the incident. Within twenty-four hours, she had increased aching, pain, and stiffness in her neck and shoulders. She saw a chiropractor and was given a diagnosis of arthritis in her neck and back, despite normal X rays. She was told to get as much rest as possible, and closely followed instructions since she was fearful of some form of progressive arthritis and total disability. She did no exercises on her own, and became almost totally inactive.

Despite her prolonged rest, her aches and pains became worse. What then followed was a series of treatments with heat and manipulation three times a week for over a year with no improvement. She visited two physicians who gave her treatment with anti-inflammatory medications and prescribed heat and neck traction to be done by a physical therapist. She worked intermittently, losing a total of about one year's work time in two years. She had no improvement in her condition. In fact, working increased her symptoms, which spread to her low back and hips, causing her concerns to increase as well.

During this time, she filed a disability claim against her employer and brought suit against the driver whose car struck hers, alleging that both her type of work and the accident caused arthritis and persistent pain. The attorney for the defendant requested consultations from various specialists, who ordered more laboratory and X-ray studies with no clarification of the problem. No report had considered the diagnosis of fibrositis. A neuropsychiatrist evaluated her and, in his report, stated very firmly that she was malingering—pretending to be ill in order to collect disability payments. Following this, she had increased difficulty sleeping, became depressed, uninterested in sexual activities, and had crying spells. Her illness created a burden for her husband, and marital stresses developed.

What began as an early and mild case of localized fibrositis accentuated by a minor automobile accident evolved into a chronic case of severe generalized fibrositis (the original diagnosis of arthritis was incorrect) and a veritable nightmare for Margie. She spent much of the two years time prior to her first visit to me under hot packs, resting in bed, visiting physicians, giving depositions, becoming frightened, angry, depressed, and suffering much humiliation.

Many other people involved in the situation—Margie's husband, her employer, the driver of the other car, and the insurance companies—suffered varying degrees of emotional upsets and economic losses as well. Bad as this scenario is, it is typical in its duration and in the tolls it takes. Above all, it is typical in its effect on Margie's fibrositis: Her discomfort worsened as the situation played itself out. In a very real sense, the illness fed on the turmoil it generated.

To physicians skilled in treating musculoskeletal disorders, and fibrositis in particular, it is all too apparent that early diagnosis and treatment—plus large doses of explanation and reassurance—would have led to a rapid and happier resolution of Margie's fibrositis, saving her and those around her much grief and economic loss. At the point when I first saw her, however, it was also apparent that in such an emotionally charged atmosphere, Margie's rehabilitation would not begin until all of the litigation she was involved in was settled, and this indeed proved to be the case.

PUTTING IT ALL TOGETHER

In the workplace, we find that persistent and chronic localized fibrositis is usually not only a physical condition of the muscles stemming from muscle overuse or minor trauma, and for the most part it will not respond simply to medication or some other treatment prescribed by a doctor. The causes and proper treatment of localized and generalized fibrositis are understandable only in the context of the individual and the environment in which they occur.

Experiences in industry offer important insights into localized fibrositis that are applicable to generalized fibrositis as well. Multiple factors are involved in causing and perpetuating the symptoms of both forms of fibrositis. Body posture, repetitive motions, and static loading of muscles (muscle contractions without moving the associated body part); tensions, stresses, anxieties, and depression; fears and concerns generated by uncertainty when fibrositis goes undiagnosed —all these factors have a role in causing and perpetuating the pains of fibrositis. When alleged injury has occurred, disability looms, and jobs, incomes, and family security are threatened, these variables also contribute to the stress on target muscles and exacerbate the pain. In cases where prolonged and costly litigation occur, producing angry confrontations, hostilities, anxieties, and stresses, sufferers can exist in a nightmare of self-perpetuating pain.

No wonder the medical picture becomes obscured. These colliding and interrelating factors create a greater stress-producing chaos. The sufferer needs not an injection, not a psychotherapist, not a cure based on assumed but unidentified tissue damage, but a treatment program that will take all variables into account and address the production of pain at every level. The following chapter offers just such a realistic, integrated program.

CHAPTER
5

RETRAIN
and
Defeat
Fibrositis

THE TREATMENT AND PREVENTION OF FIBROSITIS

The muscular aches and pains caused by your fibrositis, be it generalized or localized, can be treated successfully. I know this because I have seen it happen time and time again, even in people who, after many years of suffering, had given up hope of any respite from this illness.

In medical circles one often hears or reads the statement that "fibrositis responds poorly to treatment." Whenever I pursue the grounds for such a statement, I learn that the speaker or writer actually means that fibrositis responds poorly to medication—an assertion that I agree with completely. In truth, medications such as anti-inflammatory drugs or pain killers provide only temporary and very limited relief from fibrositis, when indeed they are effective at all.

Your muscle miseries require an integrated, comprehensive, and holistic treatment program directed at both uncovering and treating the causes of fibrositis and eliminating its symptoms. As our knowledge of fibrositis has increased, the effectiveness of this approach has slowly become evident to physicians who treat large numbers of fibrositis patients. Whether your symptoms are those of generalized or of localized fibrositis, my RETRAIN program, detailed here, is a collection of historically proven and effective concepts, therapies, and remedies that will bring you the same relief that they have brought to millions of sufferers over the centuries and throughout the world. If you wish to learn more about these various methods as used in other contexts, see the suggested reading list at the end of the book.

As your muscle aches and pains begin to subside, you will experience an increased sense of well-being and diminished fatigue. You will also probably notice that your sleep improves and that you awaken more refreshed and rested. You may also reasonably expect to experience a distinct reduction in other symptoms frequently associated with fibrositis, such as headaches and irritable bowels.

YOU AND YOUR DOCTOR: A PARTNERSHIP

Before you begin the RETRAIN program, I must emphasize that a physical examination by a physician is necessary. As sure as you may be that your symptoms are those of fibrositis, it is wisest to make certain that you have no other underlying illness that also requires treatment.

After fibrositis is diagnosed, it is important to realize that although your physician may provide direction, you must assume an active role in your treatment program. Your relationship with your doctor must be a partnership. Once the diagnosis of fibrositis has been confirmed, your physician can offer guidance and support, but you must become the senior partner and assume the major control of your treatment. At one time or another, we have all looked to our doctors for a medicine, a pill, a potion, or a procedure that would somehow cure an illness or treat an injury with little or no effort on our part. This approach may work where antibiotic therapy or suturing up a laceration is required, but it will not work with fibrositis.

GREAT VERSUS REASONABLE EXPECTATIONS

Two questions frequently arise in my practice: How soon in the course of treatment can I expect relief of my symptoms? Can I expect my fibrositis to be cured completely?

The answer to the first question is that relief may occur within days or it may occur gradually and take many months to reach its maximum. Recall that each person comes to fibrositis by a different path, each paved with one or more obstacles to good health. The path back may be short and the obstacles easy to overcome, or it may be long and arduous, requiring a considerable amount of effort and understanding on your part. The important point is that there is a path that leads away from the muscle miseries and the other uncomfortable symptoms of fibrositis, to a feeling of good health.

The answer to the second question—Can I expect my fibrositis to be cured completely?—depends on your interpretation of the word *cured*. If you interpret it as meaning "a restoration to health or sound condition," the answer is yes. If you interpret it as meaning "to have the symptoms leave and never return," the answer becomes more complex. Fibrositis symptoms can leave and never return if you are able to deal adequately with the causes. If you do not or cannot, however, the symptoms may return. There are many parallels to this situation in medicine. For example, the diabetes mellitus that may occur with obesity can be cured by weight loss, but if the obesity is allowed to recur, the diabetes mellitus will return. A bacterial pneumonia can be cured with antibiotics, but if a reinfection occurs, the pneumonia can return.

In the case of fibrositis, it is reasonable to expect that this treatment program will lead to the disappearance or at the very least an acceptable decrease of symptoms, and that the lessons you will learn will give you sufficient knowledge to prevent those symptoms from recurring or increasing.

RETRAIN

A Rational Treatment Approach

The RETRAIN program is intended to do exactly as the name implies—to retrain both body and mind in order to reverse the muscle miseries of fibrositis. The elements of the program are these:

R – REST AND RELAXATION

E – EDUCATION

T – THERAPEUTIC MUSCLE TRAINING

R – RESPONDING TO STRESS

A – ANALGESICS AND OTHER MEDICATIONS

I – INJECTIONS

N – NEVER GIVE UP HOPE!

To be successful, any treatment program for fibrositis must be carefully individualized and tailored to your personal needs. I suggest that as you read the elements of the RETRAIN program, you thoughtfully and conscientiously consider and list the aspects of the program that bear particular relevance to your own life and life-style. You can then design a personalized program that will bring you the greatest relief from your aches, pains, and other symptoms.

R – REST AND RELAXATION

Integral to the program are periods of rest and relaxation that separate you from the mental, physical, and environmental stresses that contribute to fibrositis. These periods of calmness are vital to your eventual cure. The words *Rest* and *Relaxation* have important connotations. Rest is a period of repose, inactivity, or sleep, a relief from distresses, a peace of mind, a tranquillity, or an emotional calmness. Relaxation implies a state of loosening up, a relief of muscle tension, a release from intense concentration, hard work, or worry. The term also embraces the recreation that can bring this state about.

The R in RETRAIN can also be translated as *Recreation, Respite,* or *Rehabilitation.* All these terms imply the setting aside of a period of time to allow both mind and body to restore themselves to as close to a normal, balanced, and healthy state as possible. It is in this state that the natural healing processes of the body function best.

To be effective, the types of rest and relaxation you choose must be those you truly enjoy, preferably those you can do often and those that distract you from external stress sufficiently to allow your mind and body to achieve the intended goals. During these periods of R and R, you should avoid activities that put you in competition with others or even with yourself. These activities can be as simple as reading a good book, visiting your favorite museums, taking a walk in the park or the woods, playing noncompetitive sports, or going to a ball game. No one can tell you which is best for you; only you can make the choice.

Take a moment at this point to review your current rest and relaxation program, if indeed you have had such a program. Do you put aside time every few days to enjoy yourself, or is each event in your life just another run on the treadmill? Are most of your vacations, if you take them, restful or so highly programmed that you are more tense when you return than when you left? It's hard to find ways to relax when your responsibilities are relentless and demanding. But fibrositis is a message from your body telling you that you can no longer afford to think of rest and relaxation as luxurious. They are absolute requirements for a productive, pain-free life.

E – EDUCATION

Education is what this book is all about. It wouldn't surprise me at all if this chapter on treatment is the first one you turned to. If that is the case, stop right now, turn back to page one, and begin reading again. You cannot really appreciate or understand the treatment program without understanding this illness we call fibrositis.

The first benefit of learning all you can about fibrositis will be the allaying of your fears. Pain that is not understood will in itself generate more fear—and more pain and yet more fear. "There is no medicine for fear," as an old Scottish proverb goes—none but education. What you have learned should reassure you that fibrositis is a benign disease. It does not kill; it does not deform. It surely causes pain, but we know that its causes are treatable.

Depending on your situation, it may be enough that you alone are educated about your aches and pains. It is more likely, however, that others in your environment play a prominent role in your life—and your fibrositis. Perhaps they are family members who feel supportive but are unable to understand or help, or they may be family members who do not believe you are in pain and consider the problem all in your mind. Your own physician may be sympathetic but too busy and perhaps too ill informed about fibrositis to help. You might even be facing a skeptical employer or an insurance company examiner who turns down your claim because of a lack of familiarity with the condition. In any of these situations, the *Education* element in RETRAIN extends outside yourself to others as well.

T – THERAPEUTIC MUSCLE TRAINING

The therapeutic muscle training program outlined here has five distinct aspects:

- Muscle Stretching
- Muscle Fitness
- Massage
- Heat, Cold, and Liniments
- Alternative Therapies

The first four are classical approaches used by Western physicians, physical therapists, masseurs, and masseuses to treat muscular aches and pains. The fifth, alternative therapies, involves other effective methods of muscular training and therapy. Of these, some have developed relatively recently, and others have been used for many centuries, primarily by Eastern cultures, to induce calmness, tranquillity, coordination, and a harmonious relation between mind and body.

Before going into the specifics of the therapeutic muscle training program, it will be useful to review some facts about muscles. Our muscles constitute the largest organ system in the body; they make up as much as 40 percent of the body weight in an adult male. The muscles that control movement in our bones and joints are called *voluntary muscles*, because they are under our voluntary control. This is an important point, since many other organ systems that may be the targets of stress, such as the heart or intestines, are for the most part beyond our conscious control. Thus, we are more capable of changing our muscles than these other organs. We can—by design—weaken or strengthen our muscles, flex or relax them, and rest them or drive them to the point of fatigue, exhaustion, and pain. We can use them or abuse them. In short, here we have many choices and much control.

One hundred fifty years of investigation have shown that the muscles of fibrositis sufferers are normal. That is, they show no consistent biological, biochemical, or physiologic abnormality. In addition, there is no evidence to indicate that reasonable muscle stretching, physical fitness exercise, or massage will cause damage to a normal or a fibrositic muscle.

With these facts in mind, we can turn our attention to the specific kinds of muscle use that predispose susceptible people to the aches, pains, and stiffness of fibrositis. We know from the preceding chapters that emotional stress, posture, or work causing prolonged and unrelieved muscle contractions can be the source. We know that frequent repetitive motions performed by a hand, arm, or shoulder on one side can also be the villain. And we know from recent studies that in general fibrositis sufferers are below normal in their state of muscular physical fitness.[1,2]

Here, an obvious question arises: Can one relieve the symptoms of fibrositis by increasing the fitness of the muscles? The answer, as confirmed by empirical findings, is a resounding yes. In one recent study, Canadian physicians evaluated the results of a supervised cardiovascular-fitness training program given to fibrositis patients. They found an increase in the work capacity of the participants and a decrease in chronic muscle aches and pains.[3] I personally have noted an interesting phenomenon in many fibrositis patients in this regard. Before their aches and pains began, they had been active, physically fit, and feeling great. Then, for various reasons—a minor injury, a transient illness, a new job that was sedentary and time consuming—

their physical activities dramatically decreased and the muscle aches and pains of fibrositis followed. Both these people and their physician treated their aches and pains as symptoms of an undetermined disorder requiring *increased* rest and *decreased* physical activity. This approach only exacerbated the symptoms. After the proper diagnosis of fibrositis, plus doses of explanation, reassurance, and a return to the physical activities these patients enjoyed, the fibrositis symptoms rapidly decreased or disappeared.

Muscle stretching is another time-honored method of improving muscle function and inducing muscle relaxation. Here we can learn a lesson by simply following some good athletic practices. If you observe an athlete, you will find that he or she may warm up by performing various stretching exercises before practice or competition. Afterward, the athlete will usually stretch again in order to help muscles relax. You don't have to be a competitive athlete to gain the benefits from stretching. You can profit by taking your own seventh-inning stretches on a frequent basis.

Another wonderfully soothing way of getting those aching muscles to relax is through massage. Here the muscles are treated to a hands-on form of therapy that can help to interrupt the circle of tension, spasm, and pain fibrositis sufferers endure.

Finally, applying heat, cold, or liniments to the skin over the sore muscles can give temporary relaxation and relief.

Physical Therapy—Roles and Goals
If you visit your physician with chronic fibrositic muscle aches and pains, there is a high probability that he or she will refer you to a physical therapist as part of

your treatment. This is a reasonable approach, and if you are uncertain as to your needs, it may be an essential one. With this in mind, it's appropriate to consider both the therapist's role and the goals you should both be aiming for.

A physical therapist is a health professional who has been trained in physical rehabilitation, the various modalities that are used to restore and maintain joint and muscle function in diseases or illnesses such as arthritis and rheumatism, strokes, and neurological conditions, or in injuries. These treatments may include the application of heat or cold, exercises, improving muscle and joint mobility, and joint and muscular pain relief. Many therapists are also well versed in maintaining and improving cardiovascular fitness.

Where the therapist is knowledgeable about fibrositis, he or she can be of great help in setting up a treatment program that you can eventually carry on by yourself. That is the ideal, but your actual situation may be less than ideal. The diagnosis may be incorrect, for example, as when fibrositis is mistaken for generalized arthritis. The physical therapist could outline a comprehensive program, but he or she might simply offer heat and massage. The physical therapist may know much about fibrositis or may never have heard about it. If your situation is less than ideal, you could return to the physical therapist time after time, spending hours receiving hot packs, ultrasound, and diathermy treatments, plus a dab of massage. Each such session might give you some relief for an hour or so, but in the long run the results would be frustrating to you as well as to your physician and therapist.

When your therapist does not have a basic understanding of fibrositis and its causes, a good first step would be to have him or her read this book. In that way, both of you can become informed partners in setting the goal of an individualized muscular treatment program based on your needs that you can eventually carry out on your own.

One practical economic note about physical therapy: Most health insurance policies will reimburse you for a "reasonable" number of visits to a licensed physical therapist but will question the need if the therapy sessions go on month after month (read the fine print of your policy). Your health insurance policy will probably not reimburse you for visits to masseurs or masseuses or to those who teach or give what is often called alternative therapy, such as yoga, acupuncture, and the like. This is unfortunate but a reality, so keep it in mind, particularly if finances are a problem for you.

Muscle Stretching

The stretching postures that follow are designed to help relieve the tensions that build in our muscles during our everyday activities. The goal of STRETCH-INGS is to provide a simple method of periodically offering our muscles a chance to relax and prevent the vicious circle of muscle tension, aching, pain, and stiffness. The exercises incorporate features of the yoga postures known as *asanas,* and the yoga principle of the regulation of breathing known as *pranayama.*

The exercises can, and should, be done often during the day and particularly at those times when you feel the need for that seventh-inning stretch. With the full recognition that it is often very impractical to change clothes every time you feel the need to stretch, they were designed to allow you to do them while fully dressed and in a minimum amount of space. They require only about ten minutes per session, and should be done two to three times a day, or even more often if you are sedentary for more than two hours at a time. This is a particularly important point, especially if you spend long hours at a computer, for example, or are involved in other tasks requiring repetitive actions or static muscle loading.

The STRETCHINGS group consists of ten stretches and focuses on those muscles that are most likely to be involved in muscle tension states and fibrositis—those in the neck, shoulder, back, hip, and leg areas. They are also very effective in helping to loosen the muscles before and after doing aerobic or muscle strengthening exercises.

Important Principles

There are two basic principles to keep in mind while doing STRETCHINGS:

The first is that you shouldn't stop breathing. Now that might seem quite obvious to you; however, many people do tend to hold their breath unintentionally, especially when they have not had much experience with stretching and posturing. Breath holding detracts considerably from the effectiveness of the exercises by reducing the oxygen content of the blood. Our muscles need oxygen in order to function. When the oxygen in the blood supply to the muscles is decreased below a certain point, the muscles begin to exhibit some very uncomfortable behavior. They first begin to fatigue more rapidly than normal and then may begin to cramp. In addition, prolonged breath holding can cause some lightheadedness. Breath holding also detracts from the inner tranquillity that can result from properly performed stretches. Breathing is best done through the nostrils and not the mouth. The breathing should be slow, deep, relaxed, and deliberate.

The second principle is that your muscles should be stretched gradually. Avoid the urge to bounce or jerk your muscles back and forth during stretching. It is best to stretch your muscles during exhalation, then maintain the posture during inhalation, then gradually stretch again during exhalation, doing this repetitively. This will allow you to stretch your muscles to their fullest capability at the time. The key words in stretching are *slow and gentle*.

If a mirror is available, it is helpful to do the STRETCHINGS while observing your stance and posture. Each stretch should embody at least six inhalations and exhalations, although only two or three can suffice in the early weeks when you begin the stretching postures. You should certainly do more as you become more comfortable with the movements. Now— let's begin.

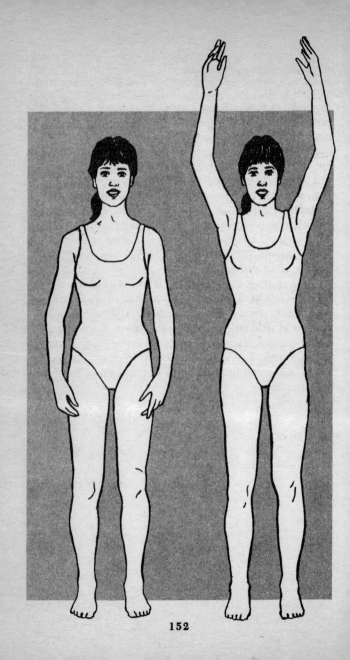

STRETCH

Figures 1 and **2**–*Assume the following* basic position. *Stand comfortably with your feet directly below your hips and shoulders and your toes pointing forward. Distribute your weight equally on the balls and heels of your feet. Keep your legs straight and the front thigh muscles tightened slightly so that the kneecaps rise. Flatten your back by slightly tilting your tailbone forward (in a pelvic tilt). This will flatten out your back. Keep your palms at your sides. Hold this position for about thirty seconds. Relax your upper body totally; look straight ahead. Breathe normally and gently.*

For the stretch, reach up with both arms, and hold your arms in a comfortable position. For most people, the elbows will be slightly bent; for others, the arms will be straight up. Stretch for the sky as you inhale. Imagine that your spine is lengthening as you exhale. Inhale, and stretch for the sky again. Repeat this at least six times, and as many times as is comfortable. Be sure to keep looking straight ahead. Now lower your arms and place your hands at your sides.

THE HUG

Figure 3—*Assume the basic position as in the first stretch; then bend your knees very slightly. Cross your arms and put your hands on your opposite shoulders keeping your arms parallel to the floor. Your head should be in a comfortable position, preferably slightly forward. Inhale; then as you exhale bring both shoulders forward as though trying to make them touch. Repeat this. After six or more breaths and stretchings, drop your arms. Reverse the process by crossing your arms in the opposite way and place your hands on the opposite shoulders. Repeat the stretch.*

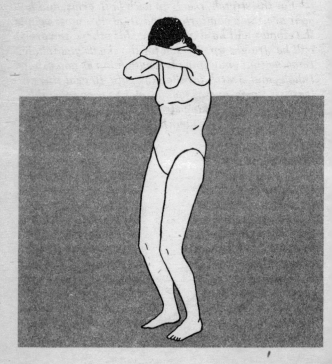

REACH FOR THE ANKLES

Figures 4 and **5**—*Again, take the basic position with your legs straight. It may be more comfortable to place your feet a little more than shoulder width apart. Keep your pelvis in the same position, and exhale as you bend your upper body at the waist to the right side, reaching your right fingers toward your right ankle. You can enhance this stretch by raising your left arm over your head and pointing your left-hand fingers toward your right ankle. Inhale, and as you exhale reach farther toward your ankle. Repeat this six times and return to the basic position. Repeat the same sequence but this time to the left. Return to the basic position.*

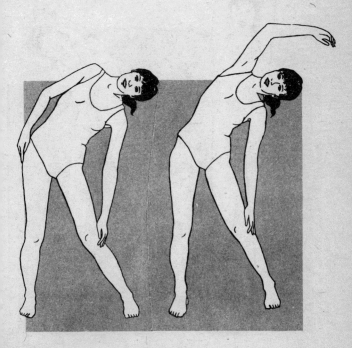

EXTEND THE SHOULDER

Figures 6 and **7**–*Assume the basic position and place your left hand behind your head and your left palm on your right shoulder blade. Grasp your left elbow with your right hand, and while exhaling gently, bend your upper body to the right, further extending the shoulder and stretching the muscles. Inhale; then exhale and stretch. After six or more stretches, return to the basic position and repeat the sequence on the opposite side.*

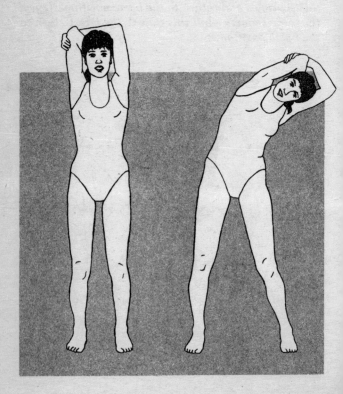

TWIST

Figures 8 and **9**–*Assume the basic position and
cross your arms and hold on to your opposite elbows.
Keep your arms parallel to the floor at shoulder level
and your head in a comfortable position. Inhale.
Then, as you exhale, move only the portion of your
body from the waist up and twist to the right. Inhale;
then while exhaling, twist slightly more to the right.
After six or more breaths, return to the center position and repeat the sequence to the left.*

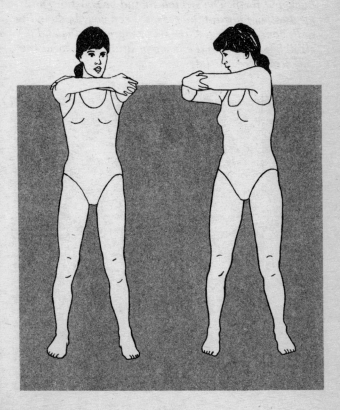

CALF STRETCH

Figure 10–*Stand about twelve to fifteen inches in front of a wall or other solid structure, such as a high countertop. Place your palms out for support, and move your right leg behind you, keeping both feet solidly on the floor and your toes pointed forward. Your left foot will not move, but your left knee will bend. With each exhalation, stretch your right foot slightly farther backward and into the right heel, putting more stretch on the right calf muscles. Do not bounce back and forth. The stretch should be constant, with gradual, not abrupt, increases. Keep your body in line with your right leg. After six or more breaths and stretches, return to the basic position and repeat the calf stretch on the left.*

HOLD THE KNEE

Figures 11 and **12**—*Stand in the basic position with your left shoulder about a foot from a wall or other support. Place your left hand on the wall for support, lift your right knee toward your chest, and grasp the knee with your right hand. With each exhalation, draw the knee closer to your chest, and tighten the abdominal muscles. Maintain the rest of your body in the basic position with the pelvic tilt. After six breaths, lower the leg, and do exercise I (Instep to the rear). Repeat exercises H and I with the other leg.*

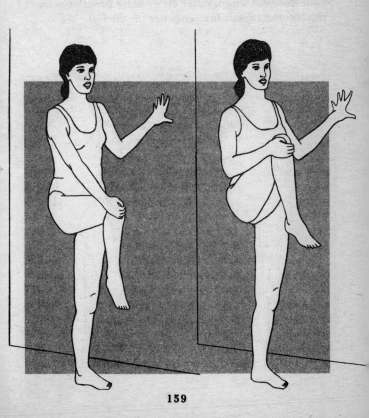

INSTEP TO THE REAR

Figures 13 and **14**–*Caution: If you have a knee problem, avoid this exercise. Begin with the same position as in the last stretch. Bend the right knee and grasp the front part of your right ankle with your right hand. As you exhale, pull the ankle toward your right buttock. You will feel a strong pull on your front thigh muscles (the quadriceps). Do not let the knee drift outward. It is important in this exercise to maintain the pelvic tilt (keeping your back flat); otherwise you will arch your back and decrease the stretch in your anterior thigh. After six or more breaths, release the leg and repeat the sequence on the left.*

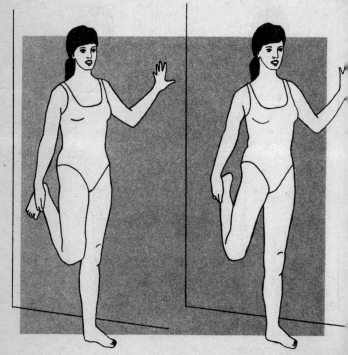

NECK FORWARD

Figures 15 and **16**—*Begin with the basic position. Move your hands behind you and intertwine your fingers with the palms facing each other. Slowly lean forward while exhaling, keeping your back flat and bending from your hip joints. If this is uncomfortable, round your back slightly and bend your knees. Keep your abdominal muscles tight. You will begin to feel the stretch in the back of your thighs in what are called the hamstring muscles. With each exhalation, bend slightly more forward at the hip joints. If your low back feels strained, bend only to a comfortable position. After six or more breaths, slowly come back to the upright position while inhaling.*

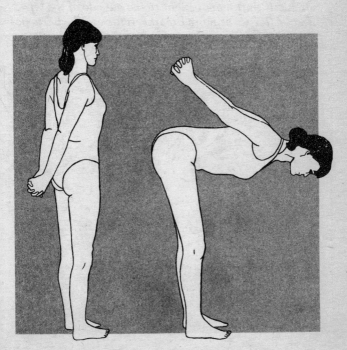

GROIN STRETCH

Figure 17–*Begin with the basic position, standing with the outside of your left foot parallel to and directly against a wall. Step away from the wall with your right foot, pointing your right toes away from the wall. At this point, your right and left feet will form a 90-degree angle. Bend the right knee to form a right angle as well. If the position feels strained, you can support your weight by placing your right forearm or both hands on your right knee. Tuck your tailbone forward. You should feel the maximum stretch in your left groin and inner thigh areas. Now, with the left leg straight, rotate the left knee slightly to the wall. This will give added stretch to the groin muscles and will shift more weight to the outside of the left foot. After six or more breaths, return to the basic position, then repeat the sequence using the opposite leg.*

STRAIGHT STRETCH

—*Repeat the first stretch.*

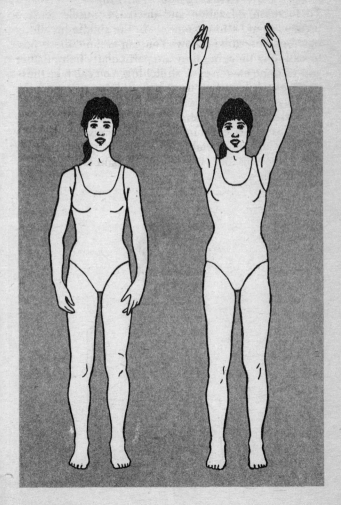

Relaxing and Breathing

To increase relaxation and decrease muscle tension further, spend a few moments in the simple breathing postures described below. You can assume these postures at any time, but they are particularly helpful after you've done exercises or stretching. You can take these postures while sitting in a chair or lying on your back on a rug or mat.

First, place both palms on your upper chest (Figure 18). Close your eyes and relax your muscles. Breathe normally and comfortably. As you inhale, imagine your breath moving into your hands. Remain for at least six breath cycles.

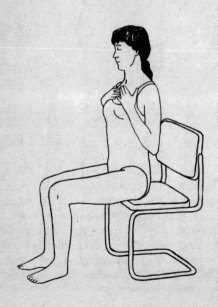

Repeat the breathing posture with your hands on your lower ribs. Repeat again with one hand on the chest above the sternum and the other on the lower abdomen just above the pubic area (Figures 19 and 20).

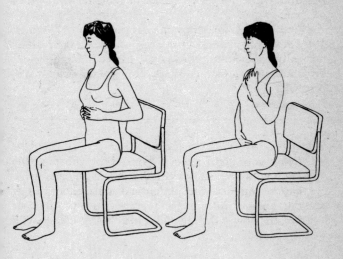

Abdominals

Our abdominal muscles play an important part in helping us to maintain proper posture and avoid muscle imbalances in the upper and lower back. The following exercises will help to strengthen your abdominal muscles, stretch your back muscles, and improve your posture. You can combine these exercises with the STRETCHINGS program, but you may want to don more relaxed clothing.

Lie flat on your back on a rug or soft mat with your knees bent. Tighten your buttocks slightly and tilt the upper part of your pelvis backward (in the "pelvic tilt") to flatten the small of the back. This is the *basic position* for the following abdominal exercises (Figure 21).

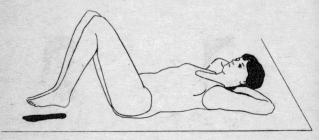

1. Draw your right knee to your chest, grasping the knee with your hands (Figure 22). With each exhalation, contract your abdominal muscles and draw the knee a little closer to your chest. After six breaths, lower the leg and repeat the sequence with your left leg.

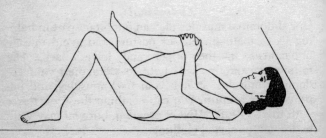

2. Draw both knees to your chest and grasp them with your hands (Figure 23). Draw the knees closer with each exhalation as you contract the abdominal muscles, as in the previous exercise. After six breaths, lower the legs to the floor.

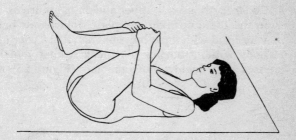

3. Place your hands behind your head with your fingers entwined. Your elbows should touch the mat or rug. Maintaining the pelvic tilt and keeping your knees together, bring both knees toward your chest (Figure 24). Then lower your legs until your feet are

one or two inches above the floor. Exhale as you bring your knees to your chest, and inhale when you lower your legs. This exercise will help strengthen your lower abdominal muscles. Start with ten repetitions and increase as your muscles get stronger.

4. To strengthen your upper abdominal muscles, start in the basic position and place your hands behind your head with the elbows out flat. Raise your upper body without flexing your neck so that your head rises a few inches above the floor (Figure 25). Keep

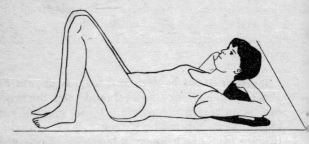

your forearms parallel to the floor—don't bring the elbows toward each other. With your shoulders and neck relaxed, lower your upper body until your head almost touches the floor. Exhale as you raise your upper body and contract your abdominal muscles; then inhale while lowering it. Repeat the sequence as in the previous exercise.

5. To stretch and strengthen the oblique abdominal muscles, get into the basic position, with your hands behind your head. With your feet on the floor and

your knees bent, move both knees to the right and your head to the left as you exhale (Figure 26). The

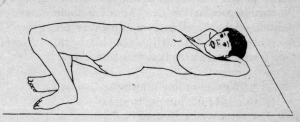

elbows should touch the floor. Inhale. Then, as you exhale, stretch your knees more to the right and your head to the left. After six breaths, return your legs and head to the center and repeat the sequence on the opposite side.

6. Repeat the first and second abdominal exercises.

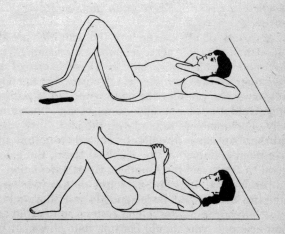

Muscular Fitness

Over the past twenty years, there has been an increasing emphasis in this country on the benefits of exercise for cardiovascular fitness. There are some other less-publicized but well-known pluses to being fit. The American Heart Association's booklet entitled *Exercise and Your Heart* states that regular exercise:

- gives you more energy
- helps in coping with stress
- improves self-image
- increases resistance to fatigue
- helps counter anxiety and depression
- helps you to relax and feel less tense
- improves the ability to fall asleep quickly and sleep well
- tones your muscles

I could have been reading a pamphlet entitled *The Beneficial Effects of Exercise on the Symptoms of Fibrositis!*

A variety of brisk exercises and activities are designed to help you achieve fitness by making your heart and muscles work harder. These exercises are often referred to as *aerobic,* which means that the body uses oxygen to produce the energy needed to perform them. Guidelines for aerobic exercises have been gradually developed and refined over the past thirty years, and there are some excellent standards to follow that have been endorsed by the National Heart, Lung, and Blood Institute and the American Heart Association.

Aerobic activities and exercises do get your heart pumping harder and faster and do put some added stress on your musculoskeletal system, so to avoid any complications, you should consult your physician before embarking on any such program. This is particularly important advice if you have or suspect you might have high blood pressure, heart problems, or arthritis.

You can attain fitness by:

- exercising at least three times a week
- raising your heart rate to between 60 and 75 percent of your maximal rate (defined below)
- maintaining this heart rate uninterrupted for at least fifteen to thirty minutes

The fibrositis sufferer should begin the program gradually. Most people with fibrositis tend to enter a program with a vengeance: "Let's do it all at once." My advice is don't. Go slowly, and allow your muscles to achieve fitness gradually. Also, if your muscles ache a little more early in the exercise program, don't worry about it; *work through it*. Mild aching following exercise is a normal phenomenon. You are not going to do any harm to your muscles with a reasonable exercise program.

If you plan to exercise three times a week, pick one or more exercises you enjoy. If you don't like what you're doing, you probably won't stick to it for very long. The list of activities you can do briskly is long enough to provide you with some choices—jogging, running in place, hiking uphill, jumping rope, rowing, stationary cycling, bicycling, and dancing or doing calisthenics to music, to name a few. A recent article in the *Journal of the American Medical Association* pointed out that brisk and consistent walking can produce heart rates in the fitness-training range.[5] The article also noted that walking was by far the most common exercise activity recommended by physicians.

If you have arthritis as well as fibrositis, don't despair—there are many ways of doing nonimpact aerobics that will still get you into good condition while protecting your joints. For further information on programs available in your area, contact your local Arthritis Foundation.

To reach the goal of getting your heart rate to 60 to 75 percent of your maximal rate, you have to know what your maximal rate should be. A simple rule of thumb is to subtract your age from 220. For example, if your age is 50 years, your maximal rate should be 170 heart beats per minute. Calculating 60 to 75 percent of that gives you a goal of 102 to 127 beats per minute. To check your heart rate, count your pulse for ten seconds and multiply by six, or count for fifteen seconds and multiply by four. Exercising above 75 percent of your maximal heart rate should be reserved for those in excellent physical condition. For your convenience, the following table lists the target zones by age:

Age	Target Zone (60–75%)	Average Maximum Heart Rate
20 years	120–150 beats per min.	200
25 years	117–146 beats per min.	195
30 years	114–142 beats per min.	190
35 years	111–138 beats per min.	185
40 years	108–135 beats per min.	180
45 years	105–131 beats per min.	175
50 years	102–127 beats per min.	170
55 years	99–123 beats per min.	165
60 years	96–120 beats per min.	160
65 years	93–116 beats per min.	155
70 years	90–113 beats per min.	150

Don't forget that your goal is exercising enough to maintain a minimum of fifteen to thirty minutes in your target-range heartbeat. Also, remember to do some muscle stretchings before and after your aerobics.

Massage

Massage is the earliest recorded form of physical therapy, having been used by the Chinese more than 3,000 years ago and by Hippocrates 2,400 years ago. During the late 1800s, massage was the major form of therapy for fibrositis, and it still has an important place in its treatment. The reason for its popularity throughout the ages is very straightforward—it works. Massage is soothing, relaxing, rejuvenating, and pleasant. In addition, it is wonderfully safe. It has no dangerous side effects, you won't become allergic to it, and it is not addicting.

Massage alone is not a cure, but when properly done, it can afford temporary relief to your aching muscles and assist in decreasing the pain and muscle tension that characterize fibrositis. A thorough massage will take close to an hour. The effectiveness of massage will depend not only on the experience of the masseur or masseuse, but, to a considerable degree, on your willingness to "flow" with it. A good masseur or masseuse can immediately detect any muscle tension or unwillingness to be touched and will attempt to put you at ease. The secret of getting a good massage is to relax and allow your muscles to be as limp as possible.

There are three basic forms of hand manipulation used. The first is light or hard stroking, which improves blood flow through the muscles and helps them to relax. The second is compression, which includes kneading, squeezing, and friction. This is helpful in stretching the muscles and improving their movement.

The third is percussion, in which the sides of the hands strike the muscles, again improving circulation. Another technique often used by physical therapists is vibration, applied through the use of electrical vibrating massagers. All types of massage are variations on those three themes.

The most popular form of massage given in the United States, often called the *Swedish massage,* employs stroking primarily, with some compression and percussion. Oils are used to facilitate the smooth stroking of the hands over the skin. Among the many used are mineral oil, safflower oil, avocado oil, and other types of vegetable oil. Many practitioners add various scents to the oils, which can bring another relaxing dimension to the massage, since pleasant aromas can in themselves be relaxing. My personal favorite oil is a nonsticky almond one with an added light almond scent. If you have never had a massage, I suggest starting with the Swedish form. It can be done very gently if your muscles are tender and very firmly after you improve.

Another form of massage that is gaining popularity in this country is the Japanese technique known as *shiatsu.* The word comes from *shi,* meaning finger, and *atsu,* meaning pressure. As the name implies, the fingers (and even the palms, elbows, and knees) are used to bring pressure on the muscles. Shiatsu is related to the technique of *acupressure* (as opposed to *acupuncture,* in which needles penetrate the skin and muscles). Many of my patients are ardent devotees of this type of massage. Shiatsu probably should not be the first type of massage that you get, however. The pressures used often cause some degree of discomfort initially, although the subsequent feelings can be a sensation of pervasive and intense relaxation and pain relief.

Reflexology is another variation on the compression technique, generally performed on the soles of the feet. The theory is that the muscles and other organs of the body have neural arcs and representations that can be mapped out on the soles or the palms, and that manipulation of these areas will improve the function of specific muscles and other organs of the body. Reflexology bears a similarity to the shiatsu technique but is based on Western rather than Eastern concepts of innervation and energy flows. Although the massage is usually given to the feet alone, the overall bodily feeling of muscle relaxation can be surprisingly effective. The theory may be open to question, but the results are not. All I can say is try it, and I predict that you'll like it.

For various reasons, it may not be convenient for you to leave home and get a massage when you most need it. Don't despair; there are other approaches that you can take. If you have a spouse or friend who is willing to learn massage, that can be one answer. Another is to find a good shiatsu teacher who will instruct you in some finger-pressure techniques that you can employ on yourself. Yet another is to get an electric massager, found at many medical supply houses, sporting goods stores, or department stores. With this instrument, it is relatively easy to give yourself a massage on your neck, upper back, arms, and legs. It may not be as relaxing as a massage by someone else, but it will be far better than nothing.

Heat, Cold, and Liniments

Of heat and cold, heat is by far the choice of most fibrositis sufferers for muscle relaxation and relief of aches, pains, and stiffness. In the case of acute strains or sprains, however, the application of cold can be beneficial.

You can apply heat through a variety of methods—by taking a hot shower; soaking in a bathtub, hot tub, or Jacuzzi; or using an electric heating pad. Physical therapists often use hot packs, and you can purchase these yourself at orthopedic supply houses. These packs are made to be soaked in hot water and to hold the heat for a considerable period of time. For those interested in the latest devices, there are even packs available that can be warmed up in a microwave oven! With all these forms of heat, the trick is to be cautious and not burn yourself—it happens all too often. People get very relaxed, fall asleep, and allow their skin to overheat.

Cold too can be applied in various ways. You can put some ice cubes in a plastic or rubber container, wrap it with a towel, and apply it to the painful areas. You can buy packs, such as Blue Ice, in sporting goods stores and orthopedic supply houses that are designed to be frozen in your freezer. Never apply these to your skin directly, however; rather, wrap them in a towel or some similar insulation. You don't want to freeze and damage your skin; you just want to cool it and the underlying muscles.

A liniment is a medicated liquid that is rubbed on the skin in order to relieve pain in deeper tissues by counterirritation. By irritating the skin, it decreases pain perception in other areas, theoretically by closing the "pain gate." Methyl salicylate—also known as oil of wintergreen, sweet-birch oil, gaultheria oil, or betula oil—is the most widely used substance in the many available liniments. It has the distinctive aroma of wintergreen, which most people find pleasing, and is not absorbed significantly into the bloodstream.

Other Avenues to Therapeutic Muscle Training

I am indebted to my patients for much of my own education about the syndrome of fibrositis, and in particular about ways to help relieve its symptoms. Many of these people, on their own, tried different alternative methods of therapy—by that I mean therapies outside the orthodox medical approach. Some methods gave them a sense of tranquillity and an improved feeling of well-being. These alternative methods may provide you with the same results.

My personal feeling about alternative methods is that if they help and do no harm, I'm all for them. The ones that people tell me they have the most success with are ones they are eventually able to do on their own. This does not surprise me, since it coincides with my perception that fibrositis sufferers do best when they assume control of their treatment.

There are many alternative methods, and a full discussion of them is beyond the scope of this book. Here I discuss three that have helped patients in my practice: yoga, *tai chi chuan*, and acupuncture.

Yoga

Mention yoga to many people, and they picture a Hindu ascetic in a loincloth whose body is contorted into a pretzellike form. That is certainly one extreme of yogic practice, but not the part of the picture that is important to us. We can choose those aspects of yoga that benefit both body and mind without adhering rigidly to the entire mystical philosophy.

Many practitioners have adapted the yoga postures and breathing exercises to Western concepts and practices. Their major emphasis has been on two of the eight stages—the therapeutic, relaxing, and bracing effects of the yoga postures (*asanas*) and the breathing exercises (*pranayamas*), with their ability to help free the mind and to control inner experiences in a positive way.

If this approach sounds appealing to you, I suggest you find someone who teaches a course in yoga and ask if you can observe a session before deciding to try it yourself. There are many schools of yoga scattered around the country, and even health clubs, colleges, and city recreational departments offer yoga programs. It is possible to learn something from reading books on yoga or following along with television programs, but you will benefit more from someone who can guide you on a more personal level.

Tai Chi Chuan

Tai chi chuan has been described as "body and mind in harmony." Its origins lie in the Chinese philosophies of the causation and treatment of illness and the maintenance of a healthy mind and body that were developed over three millennia. The concept of *tai chi chuan* itself dates back to about A.D. 1000. It is a form of moving meditation based upon our place in nature, the movement of animals, and traditional fighting postures. The movements are dancelike and emphasize both mental and physical coordination. This is a "soft exercise" that can be done by almost anyone, since it does not require a great amount of energy and does not put undue stress on the heart, muscles, or joints. You may well have seen it being performed without being aware of what it was. Many television news reports and documentaries on China show hundreds and even thousands of people engaged in *tai chi chuan* in city squares.

Tai chi chuan is rapidly gaining acceptance in this country, and instruction in it is usually easy to find. Again, if this form of exercise interests you, I suggest you contact your local health clubs, colleges, or city recreational department for information on classes available in your area.

Acupuncture

Of all the questions I've been asked about treating musculoskeletal illnesses with alternative therapies, the greatest number have focused on acupuncture. The Chinese theory of acupuncture relates to a complex balance of cosmic forces of life energy and disease involving the *yin*, the female principle, the *yang*, the male principle, and twelve channels throughout the body.[6] These channels, six of which belong to the yin principle and six to the yang, are thought to be deeply embedded in the muscles. But at least three-hundred-and-sixty-five points emerge close to the surface, where they can be needled. Long thin needles of various metals inserted at these points are thought to correct any imbalances of yin and yang and thus to return the body to a healthy state.

This theory bears little relationship to our Western concepts of disease and illness, but as we know, theory and results don't always match. In the context of fibrositis, the question is, Does it work? Here, again, I cite the experience of my many patients who have had acupuncture treatments. Overall, the results have been very disappointing as compared with those obtained with yoga and *tai chi chuan.* I have no objection if my patients wish to try acupuncture, but I can't recommend it with a great degree of optimism.

R – RESPONDING TO STRESS

For most people, the word *stress* conjures up a picture of something bad, something to be avoided. That's true in part: As this book has pointed out, stress can create and contribute to our bodies' illnesses. But there are good sides to stress that are often overlooked. Stress can help us grow and develop. It can temper us to better withstand the challenges of life, as extremes of heat and cold can temper iron into steel. It can be the catalyst for innovative and progressive ideas.

It is crucial to remember, in determining what effects stresses can have on our minds and bodies, that it is not so much the type of stress but how you respond to it. That's important, especially where fibrositis is concerned. How you respond to and deal with the ever-present stresses of life can make considerable differences in your symptoms—for the better or for the worse. A reasonable step to take in moderating your responses, then, would be to become familiar with your own stress-induced reactions and their end results.

Recognizing How Stresses Affect You

What is needed is some honest introspection. Start with a list of those situations you know make your fibrositis symptoms worse. For example,

- If you do not do a task "perfectly," do you become anxious and achey?
- Are you upset and tense when anyone in your family or at the job leaves a task undone?
- Does criticism depress you and make you feel miserable?

- Is a member of your family or your boss a pain in the neck (or some other part of the anatomy)?
- Do your muscles hurt more when you're on the job than when you are on vacation?

Answers to such questions will point most urgently to where you need to change—to alter your response for the sake of your well-being.

Next, list those situations you recognize as stressful to you but that do not seem to coincide with a worsening of your symptoms. For example,

- Do you feel guilty when you do not assume a task that a family member has left unfinished?
- Do you hate your boss or your job?
- Are you constantly anxious and worried about finances or what tomorrow might bring?
- Do you feel worthless or depressed?

Do not be fooled. Despite the seeming lack of any one-on-one relationship to your symptoms, these responses are ingredients in the medium on which your fibrositis thrives. They, too, point to areas in which appropriate changes in your responses will help to relieve your muscular miseries.

Finally, list those stressful situations that make you feel better. For example,

- Does the challenge of a new job excite and exhilarate you?
- Do you feel more relaxed after competing in a tennis tournament—win or lose?
- Do you feel calmer and more self-assured

when you can successfully talk out a problem with a family member, friend, or coworker?
- If you are tense speaking before groups, does each experience tend to lessen that tension and increase your self-esteem?

By candidly preparing these lists, you can draw your own stress portrait—a picture of your internally and externally generated stressors. This can serve as the framework for your personalized stress-management program. It will also point to the many methods you have used and can use again to make stress work for you, not against you.

Some Methods of Dealing with Stress

Your responses to stress, like your personality and susceptibility to illnesses, are determined in part by your nature, or genetic makeup, and in part by your nurture, the environment in which you were raised. You cannot change your genetic background, but you can make significant changes in your learned responses that can improve your physical and mental health and change for the better the way stress affects you.

Two of the first three elements of the RETRAIN program—rest and relaxation, and therapeutic muscle training—are basic and integral parts of any stress-reduction program. To these you can add a variety of effective methods, ranging from the very old to the very new. It is up to you to choose those methods that are most compatible with your personality and lifestyle and those that you can actually apply.

Talking It Out

It is often helpful to discuss your emotions and concerns with a sympathetic person, such as a spouse, friend, or clergyman. This will help you let off a little steam, defuse some anger, relieve tensions, and give you a different perspective on and approach to your problems. In many cases, this may be more effective for you than professional counseling.

Where specific stresses are in place—for instance, an alcoholic spouse or teenager, a family member who is on drugs, a family member who is a compulsive gambler, an abusive husband, or a loss from suicide—the specialized support group meetings held in almost every city can be extremely helpful. Your local newspaper or telephone book is a good place to find out about these various groups. A very brief listing includes Al-Anon and Alateen for family and friends of alcoholics, Toughlove for family and friends of children and teenagers who have drug or alcohol problems, Abused Women's Services, Gamblers' Anonymous (Gam-Anon), Debtors' Anonymous, Suicide Loss, and AIDS support groups. The growing number of such groups in this country attests to the fact that they fill a large void in many people's lives. Many of my patients have told me that they have found much solace and relief from anxiety, tension, and depression in support groups and have learned, through the experience of others who have successfully dealt with the issues they faced, how to handle the stresses in their lives.

A Balanced Sense of Commitment

Many writers on stress reduction emphasize the need for a sense of commitment—a need for realizing achievable goals. Fibrositis sufferers, however, do not necessarily lack commitment: Rather, they often have an overbearing sense of commitment, not only to their families but to their coworkers, their neighbors, and to society as a whole. They neglect no one except themselves. The thought behind this wholehearted effort to do and achieve may be altruistic, but the results can be great amounts of unwanted stress and pain.

What is needed in such cases is a balanced sense of commitment—a realization that the giver as well as the receiver needs care and attention. If this picture of a perpetual giver matches your image of yourself, remember that your emotional and physical energies are not unlimited; they are as finite as everyone else's. There is a time to set priorities, a time to say no, a time to realize you cannot please everyone, and a time to demand a sharing of responsibilities.

Stress in the Workplace

If you feel that the workplace is a major source of your stress, you have plenty of company. This topic has been a growing concern among workers, union leaders, and management, and has generated a multitude of businesses offering stress-management and ergonomic programs to industry. If your employer's business is large enough to warrant the expense, and the interest is there, these programs may be available to you. If not, you will probably have to work out your own solutions.

Your success in dealing with these issues in regard to fibrositis will largely depend on how honestly and accurately you can define both the stressors and their relationship to your aches and pains. Again, you must pinpoint them exactly. Ask yourself whether the stress is due to interpersonal relationships, such as those with your boss or coworkers, or mechanical factors such as the keyboards or other tools or machines you work with or both. One pitfall common among fibrositis sufferers is the tendency to blame the job for all problems when the root of the problem actually lies elsewhere—perhaps in internally generated stress or in family strife. For people who confuse the source of their stress in this way, even changing jobs has little effect on their fibrositis symptoms.

But if indeed the workplace is the source of your stresses, you, for example, may have to direct your efforts toward improving your relationships with those you work with, taking seventh-inning stretches, learning how to operate the machines with less stress on your muscles, or physically and mentally conditioning yourself to better withstand the stressors. Quitting your job is one answer but one that could bring new, perhaps monumental stresses into your life. And such drastic measures may not be necessary if you can root out the problems and solve them through accurate assessment and commitment to *yourself.*

Humor and the Lighter Side of Life

Another time-honored method to help separate you from your stresses is to cultivate your sense of humor and allow yourself to have a little fun. Looking at the serious side of life all the time can be pretty depressing and in itself a source of stress and pain. Smiles and laughter can help you release your tensions and make the world around you look a lot happier and more hopeful.

Granted, it's hard to develop that sense of humor when you're in pain. Things may not seem all that funny at the time. One way of getting around this is to recall the times that you weren't in pain and remember what made you smile and laugh. Was it watching a funny movie or a comedy show, reading a humorous story or a good jokebook, or perhaps reminiscing with friends about some of your good old times? Try calling up memories of these happier periods in your life—times when you didn't take the world and yourself so seriously. Try taking more steps on the lighter and brighter side of life where the clouds are fewer, the sun is warmer, and the stresses of life appear smaller and more distant.

Nutrition

I'm not aware of evidence suggesting that dietary changes play a direct role in producing or improving fibrositis symptoms, but the importance of good nutrition and a balanced diet in maintaining your optimum health and improving your resistance to stress is well-known. I have seen patients who subjected themselves to rather bizarre and restricted diets in the hope of avoiding some unknown foods that might be causing their pains and fatigue. It never seemed to work. On the contrary, in many instances their health suffered because of misguided efforts. My advice is to stick to a well-balanced and healthy diet. You can't go wrong that way.

Stimulants and Depressants

If you are going to let your natural healing processes reach their maximum, I strongly urge you to avoid the use of tobacco, marijuana, and drugs like tranquilizers and pep pills. There is no place for them in a healthy body.

Alcohol may temporarily relieve some of your inhibitions and tensions, but keep in mind that it is basically a central nervous system depressant. When used very modestly—a glass of wine with dinner, for instance—alcohol can be relaxing and is unlikely to cause any significant health problems. Difficulties arise, however, when it is used in excess, or to solve or escape from life's problems and stresses. If you find that yōu are relying on alcohol at any time to help you deal with any type of situation, that is the time to stop. If you cannot, you must seek professional assistance.

Caffeine is a central nervous system stimulant. Its major natural source is coffee, but it is also found in tea and cocoa, and it has been added to many cola drinks and to medications such as aspirin. It is popularly used for its favorable effects on mood and feelings of fatigue. If used to excess, however, caffeine can increase many of the symptoms of fibrositis by producing restlessness, stress, sleeplessness, and irritability. Here again, moderation is the key word.

Imagery

To quote Mark Twain, "I am an old man and have known a great many troubles, but most of them never happened." Our imaginations, if left unchecked in a negative mode, can easily bring about trouble states involving anxiety, tension, stress, illness, and pain. Conversely, if positively directed, as with guided imagery, the imagination can be a powerful instrument for temporarily relieving these same moods and conditions.

We can use imagery to help us change the perception of our relationship to both our internal and external environments. Think of it as daydreaming in a very positive sense. In the case of fibrositis, this technique can be used as an adjunct to the other methods of treatment described. For example, if you are tense, you can quietly relax while imagining that you are on a trip far away from your problems and cares, lying by a pleasant stream and watching a beautiful sunset. And, if you have pain, you can imagine the pain leaving your body and flowing into a bottle that then drifts away with the current.

This may all sound a bit like wishful thinking or black magic, but it really is neither. Various imagery techniques to improve health have received a considerable amount of favorable attention in the last three decades both in the medical and lay communities and have become effectively integrated into many relaxation and pain-reduction programs. Although it often helps to have someone guide you in using imagery, the many available books and tapes describing various approaches to imagery techniques definitely put this methodology into the do-it-yourself category.

Meditation

Meditation, which has been practiced in many forms for millennia, is a personal and spiritual method used to induce a sense of inner peace, calm, and relaxation. It allows for a period free from the stresses and strains that surround you, and allows you to focus more fully on your deeper thoughts or religious beliefs. With the increasing appreciation of Eastern concepts of mind-body interactions, meditation has received wide attention in the West. Some of my fibrositis patients have found meditation far more effective than medication.

Relaxation and Music Tapes

Although technology has been rightly blamed for causing many of our stresses, it has also provided us with some new ways of coping with stress. One relatively inexpensive method is using a portable cassette player and music or relaxation tapes. This combination gives you many choices of stress-reduction methods and allows you to choose the time and place to apply them.

There are lots of relaxation tapes available to combat stress, and they use a wide range of techniques. Some incorporate guided or mental imagery, others self-hypnosis, and others, direct muscle-relaxation techniques. Still others emphasize the release of deleterious emotions, such as negative attitudes, anger, or guilt. Some are very broad in scope, while others concentrate on specific problems, such as pain-reduction methods, migraine headaches, or alcohol or drug addiction. Some focus on using the mind to relax the body, others on using the body to relax the mind, and still others on both.

Certain types of music can induce relaxation and bring relief from stress through the use of soothing tones, themes, rhythms, and intensities. The contemporary music often called New Age music has been composed specifically for this purpose and is so varied that it cannot be simply described. The instruments used vary from ancient stringed instruments, flutes, and Tibetan bells to the most technologically advanced electronic synthesizers. Natural sounds, such as the warbling of birds and the gentle washing of waves, are often incorporated into the compositions. The forms are equally varied, and may be based on folk melodies, classical music or jazz, or Indian ragas, or they may be ethereal and free-flowing compositions. The tempo is often slow—rarely faster than sixty to seventy beats per minute, the rate of the human heartbeat during rest. This music is best enjoyed in a quiet, peaceful setting, and it can be used to enhance imagery or accompany meditation or massage. The selection available is diverse enough to provide many to suit your own preferences and meet your needs.

Biofeedback

Biofeedback is an autoregulatory technique. Integral to this method are instruments that give you immediate and continuing information on changes in bodily functions you are usually unaware of, such as heart rate, blood pressure, extremity temperature, and basic muscle tone. Theoretically, and often in fact, this feedback can help you gain some conscious control over these generally involuntary bodily functions by making you aware of physiological changes that occur when you use the relaxation methods I have described, such as imagery, meditation, and listening to music. Some of my patients have found biofeedback to be extremely effective in helping them self-induce muscle relaxation, particularly in the case of the temporomandibular joint (TMJ) syndrome.

Counseling

There may come a time when you feel you simply can't handle all your stresses on your own—that they are beyond your control. Try as you may, you find yourself unable to shake such feelings as overwhelming anxiety, guilt, or depression. You may feel trapped in situations that are not of your own making, or in relationships that do not seem to be working out despite your best efforts. Life may not even seem worth living. At this point, you should definitely discuss your concerns with your physician and seriously consider seeking counseling with another professional, such as a family or marital counselor, psychologist, or psychiatrist.

A – ANALGESICS AND OTHER MEDICATIONS

Medications have a limited role in the treatment of both generalized and localized fibrositis. One reason is that they can alleviate the symptoms of fibrositis but don't really address themselves to the causes. Another is that very few medications have any proven value in the relief of these symptoms.

Still, medications can bring some relief of symptoms. Ideally—and in this case the ideal is often reached—fibrositis can be treated without the use of medication. From a practical standpoint, however, it is hard to justify withholding them if pain and suffering seem unbearable. The relevant questions are what helps, and what is safe?

Analgesic and Anti-Inflammatory Medications and Generalized Fibrositis

There are many nonnarcotic medications available that have both analgesic (painkilling) and anti-inflammatory effects. Aspirin and ibuprofen are available without a physician's prescription. In addition, there are various nonsteroidal anti-inflammatory (noncortisone) medications available by prescription, such as naproxen (Naprosyn), sulindac (Clinoril), piroxicam (Feldene), diclofenac (Voltaren), and salsalate (Disalcid). Acetaminophen, a nonprescription medication, is an analgesic, but has no anti-inflammatory effects. All of these agents can have undesirable side effects to some degree but are reasonably safe and known to be very effective in relieving the aches and pains of such diverse conditions as muscle strains and sprains, arthritis, tendinitis, and bursitis.

It might seem logical to assume that these medications would be of at least equal value in the treatment of the pains of generalized fibrositis. Although widely used for this, they have actually proven to be of less, and at times of no, value for it. The explanation most certainly relates to both the way these drugs relieve pain and inflammation, which is as yet unclear, and the known causes of the conditions being treated.

Strains and sprains are the result of tissue damage from injury, the pain of rheumatoid arthritis is associated with tissue damage and inflammation, and the aches and pains of influenza are associated with a viral infection. But the aches and pains of generalized fibrositis, as we have seen, are not associated with tissue damage, inflammation, or infection but are mainly the result of soma-psyche interactions generated by internal and external stress. The pains are often but one part of a picture that includes anxieties, tensions, depression, disturbed sleep, migraine headaches, and irritable bowels. It is more logical to treat these pains with agents known to ease the specific conditions and symptoms. Amitriptyline is one such medication.

Amitriptyline and Related Medications

Amitriptyline, available by prescription, is a commonly used and effective medication for many stress-related conditions and the most widely used for the treatment of fibrositis. It falls into the class of drugs known as the tricyclic antidepressants and is sold both generically and under the trade name Elavil. Its mechanism of action is not fully known; however, it may produce some of its effects by enhancing the action of serotonin in the central nervous system. It has no known direct analgesic effects when tissue damage is present but appears to affect the perception of chronic pain. Amitriptyline,

fortunately, is not addictive. It may cause some drow-
siness, so it is often prescribed as a single dose at bed-
time, the usual dose varying from ten to seventy-five
milligrams. The beneficial effects can take a few days
or a week to appear.

Amitriptyline, although often effective, is far from a
panacea. It is not universally effective for all fibrositis
sufferers, nor is it free of undesirable side effects.
These side effects include, but are not limited to, dry-
ness of the mouth, confusion, weakness, and—in high
doses—irregularities of the cardiac beat. Most fibrositis
sufferers, however, do tolerate the drug well. If not,
other medications in the same class, such as desipra-
mine or imipramine, can be tried. These drugs should
only be taken under a physician's direction and then
only in conjunction with the other treatment methods
that I have already outlined.

What about the so-called muscle relaxants? One
medication that has been of some value is cyclobenzaprine (Flexeril). Interestingly, cyclobenzaprine does
not have any direct muscle-relaxing effects, but chem-
ically it is closely related to amitriptyline and it acts
similarly through its effects on the central nervous system. It may also produce some undesirable side effects
and should only be taken under the close supervision
of a physician.

Narcotic Drugs: A Warning

Narcotic drugs, such as codeine, meperidine, morphine, and propoxyphene, are used to alleviate pain
from any source. They get the name *narcotic* from the
fact that they are capable of producing narcosis—a state
of stupor and even unconsciousness—and that they are
addictive drugs. They are fine drugs for pain that is

expected to be of short duration; however, narcotics are extremely poor choices in the treatment of fibrositis. They do not address the sources of the problem and in fact tend to obscure them and make them worse. The result is a worsening of the fibrositis plus the added burden of a drug addiction. The answer is simply not to use narcotics for treating your fibrositis aches and pains.

Medications and Localized Fibrositis

All the medications used to treat generalized fibrositis have been tried in the treatment of the localized fibrositis syndromes, usually with the same results. The anti-inflammatory medications, however, have shown some effectiveness in relieving tendinitis or bursitis when it accompanies fibrositis.

I – INJECTIONS OF TENDER POINTS

Many physicians will treat localized painful areas of fibrositis with injections of such medications as the local anesthetic lidocaine into the tender spots of the muscles. Also used are cortisone derivatives, such as prednisolone. This technique is most often applied in the treatment of the trigger points of the myofascial pain syndrome. It has also long been a standard and accepted form of therapy for other diverse forms of soft tissue rheumatism, such as the inflammation of tennis elbow or bursitis of the shoulder.

I use this procedure to treat tendinitis or bursitis pain that hasn't responded to oral medications, but I avoid injections in cases of fibrositis. To begin with, such injections don't usually help much in reducing fibrositis pain, and if they do, the relief doesn't last for long. In addition, I seriously question the advisability of injecting medications into muscle tissue that has been shown to be normal.

N – NEVER GIVE UP HOPE

An essential feature of the RETRAIN program is that you must never give up hope. This is one of the basic messages of this book. To give up hope is to relinquish the task of taming the forces that create your fibrositis symptoms—a task that only you can direct. Always remember—it will take some effort on your part to control your fibrositis, but control it you can.

As you improve, there may be periods when the pains temporarily return. This is not at all unusual. Do not be discouraged and do not give up hope. These are times not to consider the treatment program ineffective but rather to rethink and reevaluate what is happening in your life and to determine how you can respond in a more effective manner. These are times not for dejection but for remembering that you already have, if temporarily, broken the stress/pain/stress cycle and that you can break it again. Finally, these are times in which to reread the elements of the RETRAIN program and to renew your hopes, not give them up.

THE PREVENTION OF FIBROSITIS

The techniques described here in treating fibrositis can also be used to prevent it from developing. The first preventive step, however, is to cultivate the ability to recognize those situations in which it is most likely to appear. With our understanding of fibrositis increasing, we are in an excellent position to continue to evaluate the demands of the workplace and their relationship to the muscular aches and pains of fibrositis. And we can hope that the investigative tools that are being developed by researchers in biotechnology will uncover more of the roots of this increasingly pervasive illness, particularly in the case of localized fibrositis.

With regard to generalized fibrositis, early diagnosis and treatment are keys to preventing it from becoming a chronic and debilitating illness.

This stress-related disease we call fibrositis will only become a thing of the past when humankind has learned to integrate mind, body, and environment into an effectively functioning unit. Predicting to what extent this will occur in the future, or whether it will ever occur, is certainly far beyond our present abilities. Still, with the elimination of fibrositis as our goal, every step in that direction will take us toward exhilarating, productive good health.

SUGGESTED
REFERENCES
AND READING

General Information

Planetree: A Consumer Health Care Organization
2040 Webster Street
San Francisco, CA 94115
   This unique organization is an excellent starting point for past and current information about fibrositis and other medical topics, holistic medicine, and alternative therapies.

Arthritis Foundation
1314 Spring Street, NW
Atlanta, Georgia 30309
   A good source of information if you have or suspect you have arthritis as well as fibrositis. There are local chapters in many cities across the country.

American Heart Association
7320 Greenville Avenue
Dallas, Texas 75231

They, as well as their local chapters, can provide you with current guidelines to follow for your fitness program.

Muscles, Mind, and Body

Anderson, Bob. *Stretching*. New York: Distributed by Random House, Shelter Publications, 1980.

Benson, Herbert, and Klipper, Miriam Z. *The Relaxation Response*. New York: Avon, 1976.

Borysenko, Joan. *Minding the Body, Mending the Mind*. Reading, Mass.: Addison-Wesley, 1987.

Bresler, David E., and Trubo, Richard. *Free Yourself from Pain*. New York: Simon & Schuster, 1986.

Cooper, Kenneth H. *The Aerobics Program for Total Well-Being*. New York: Bantam Books, 1983.

Cousins, Norman. *Anatomy of an Illness as Perceived by the Patient*. New York: Bantam Books, 1983.

Crompton, Paul. *The T'ai Chi Workbook*. Boston: Distributed by Random House, Shambhala Publications, 1987.

Davidson, Paul. *Are You Sure It's Arthritis? A Guide to Soft Tissue Rheumatism*. New York: New American Library, 1987.

Delza, Sophia. *T'ai chi ch'uan: Body and Mind in Harmony*. Albany: State University of New York Press, 1985.

Downing, George. *The Massage Book*. New York: Random House, 1972.

Hittleman, Richard. *Yoga for Health*. New York: Random House, Ballantine Books, 1983.

Iyengar, B.K.S. *Light on Yoga.* New York: Schocken Books, 1976.

Lawrence, D. Baloti. *Massage Techniques.* New York: Putnam Publishing Group, 1986.

LeShan, Lawrence. *How to Meditate: A Guide to Self Discovery.* New York: Bantam Books, 1986.

Lidell, Lucinda. *The Book of Massage.* New York: Simon & Schuster, 1984.

Locke, Steven, and Colligan, Douglas. *The Healer Within.* New York: New American Library, 1987.

Lorig, Kate, and Fries, James F. *The Arthritis Handbook.* Reading, Mass.: Addison-Wesley, 1986.

Miller, Emmett E. *Self Imagery: Creating Your Own Good Health.* Berkeley: Celestial Arts, 1986.

Namikoshi, Tokujiro. *Shiatsu: Japanese Finger-Pressure Therapy.* Briarcliff Manor, N.Y.: Japan Publications, 1972.

Olshan, Neal H. *The Scottsdale Pain Relief Program.* New York: Ivy Books, 1988.

Ornstein, Robert, and Sobel, David. *The Healing Brain.* New York: Simon & Schuster, 1987.

Pelletier, Kenneth R. *Mind as Healer, Mind as Slayer: A Holistic Approach to Preventing Stress Disorders.* Magnolia, Mass.: Peter Smith, 1984.

Rosas, Debbie, et al. *Non-Impact Aerobics.* New York: Random House, Villard Books, 1987.

Rossman, Martin L. *Healing Yourself: A Step-by-Step Program for Better Health Through Imagery.* New York: Walker and Company, 1987.

Tisserand, Robert. *The Art of Aromatherapy.* Essex, Eng.: C. W. Daniel, 1987.

Tobias, Maxine, and Stewart, Mary. *Stretch and Relax.* Tucson, Ariz.: Body Press, 1985.

Veith, Ilza, trans. *The Yellow Emperor's Classic of In-*

ternal Medicine. Berkeley: University of California Press, 1966.

Watson, Andrew, and Drury, Nevill. *Healing Music*. Garden City Park, New York: Avery Publishers, 1988.

Ergonomics

Human Factors Society
Box 1369
Santa Monica, California 90406

The Joyce Institute
1313 Plaza 600 Building
Seattle, Washington 98101

Gay, Kathlyn. *Ergonomics: Making Products and Places Fit People*. Hillside, New Jersey: Enslow Publishers, 1986.

Joyce, Marilyn. *Ergonomics: Humanizing the Automated Office*. Cincinnati: South-Western Publishing Co., 1989.

GLOSSARY

Acupressure The application of pressure over specific muscle sites in order to relieve pain and muscle spasm.

Acupuncture Ancient Chinese method of treating pain and illness by passing thin needles through the skin at specific points.

Adrenal gland A small gland lying above the kidney that secretes cortisone, adrenaline, and related hormones.

Adrenaline A hormone secreted by the adrenal gland that stimulates the heart and muscles.

ANA See *antinuclear antibody*.

Antinuclear Antibody One of various antibodies to components of the nucleus of human cells. They are present in many autoimmune disease states, such as systemic lupus erythematosus, but are also found in about 5 percent of normal people.

Anxiety A state of being apprehensive, worried, or concerned about what may happen in the future.

Arthralgia Pain in the joints in the absence of arthritis.

Arthritis From the Greek word *arthrond* meaning joint, and the suffix *itis*, meaning inflammation. It generally means damage to a joint from any cause, such as infection, trauma, or inflammation.

Asana Yoga posture. The third stage of yoga.

Autoimmune Disorder An illness resulting from the production of autoantibodies that attack and damage tissues of the body.

Bursa A small sac containing a sticky fluid that is interposed between muscles and tendons, or tendons and bony prominences.

Bursitis Inflammation of a bursa.

Candida albicans One of many yeasts or single-celled fungi. Also known as Monilia albicans. It is widely present in our environment and found normally in the human mouth and intestines.

Candidiasis An infection caused by Candida albicans in susceptible people.

Carpal Tunnel Syndrome A disorder characterized by various combinations of pain, tingling, and numbness in the palm of the hand, and often pain radiating up the forearm. The causes are many, including the repetitive strain syndrome, rheumatoid arthritis, hypothyroidism, and a host of other medical illnesses.

Cortisone A hormone that is produced by the cortex of the adrenal gland and, like related hormones such as prednisone, can be produced synthetically. It has many effects on the body, including a potent anti-inflammatory one.

Depression A state of mind characterized by marked gloominess, dejection, or sadness.

Disease A pathological condition of the body; literally, the lack of ease.

Disorder An ailment, or an abnormal health condition.

Endorphins Opiatelike natural painkillers produced in the human nervous system.

Enkephalins Similar to endorphins.

Epstein-Barr Virus The virus that causes infectious mononucleosis.

Ergonomics The study of human capabilities and limitations in relation to the work system, machine, or task, as well as the study of the physical, psychological, and social environment of the worker. Also known as human engineering.

Erythrocyte Sedimentation Rate Also called the sed rate. A laboratory test that measures the rate, in millimeters per hour, that red blood cells fall in a thin tube of blood. It is a nonspecific test in that a high rate may indicate inflammation. Two different methods are in general use; the Westergren method gives higher readings than the Wintrobe method.

ESR See *erythrocyte sedimentation rate.*

Fibromyalgia See *fibrositis.*

Fibrositis A noninfectious soft-tissue rheumatic condition characterized by muscle stiffness and pain and with no associated abnormal X-ray or laboratory findings. It is often associated with fatigue, headaches, irritable bowels, and disordered sleep.

Generalized Fibrositis Fibrositis involving widespread muscular areas.

Genetic Predisposition A susceptibility to a specific disease or illness because of a person's genetic characteristics.

Hypothyroidism A disorder resulting from a lack of thyroid hormone. Muscle aching and stiffness is often one of the characteristics.

Illness The condition of being in poor health.

Immunologic Disease A disease caused or perpetuated by an imbalance of antibody defenses; for example, systemic lupus erythematosus.

Infectious Mononucleosis A disease caused by the Epstein-Barr virus and characterized by fever, sore throat, and swollen lymph glands in the neck.

Inflammation The reaction of tissues to injury from any number of causes, such as infection or trauma. Characterized by various degrees of warmth, redness, swelling, and pain.

Internist A physician who specializes in internal medicine.

Irritable Bowel Syndrome A condition characterized by abdominal discomfort and bowel habits varying from constipation to diarrhea.

Liniments A liquid containing a medication applied to the skin.

Localized Fibrositis Fibrositis affecting localized areas of the body, such as the neck, shoulder, and arm. Also known by other terms such as repetitive strain injury or myofascial pain syndrome.

Lupus Erythematosus See *systemic lupus erythematosus*.

Massage A technique utilizing kneading, pressure, friction, and vibration, usually applied to the body to produce muscular relaxation.

Masseur A man who gives massages.

Masseuse A woman who gives massages.

Meditation Deep, continued thought and reflection.

Migraine An intense, periodically returning headache, usually limited to one side of the head.

Moniliasis See *candidiasis*.

Morbidity The rate of disease or the proportion of diseased persons in a given population.

Myalgia Tenderness or pain in the muscles. Muscular rheumatism.

Myofascial Pain Syndrome See *localized fibrositis*.

Objective Capable of being seen or measured.

Occupational Cervicobrachial Syndrome See *repetitive strain injury*.

Occupational Overuse Syndrome See *repetitive strain injury*.

Osteoarthritis A chronic disease involving the joints and characterized by destruction of cartilage and bony overgrowth.

Physiatrist A physician who specializes in the field of physical therapy and rehabilitation.

Physical Therapist A person whose profession is the practice of physical therapy.

Physical Therapy Rehabilitation concerned with restoration of function and prevention of disability following injury or disease and utilizing physical methods such as heat, cold, exercise, and massage rather than drugs.

Polymyalgia Rheumatica A condition affecting primarily those over the age of fifty and characterized by shoulder and hip girdle muscle pain and stiffness and a high sedimentation rate.

Pranayama The yoga practice of rhythmic control of breathing. The fourth stage of yoga.

Psychophysiologic Pertaining to the influences of the mind on bodily function and illness.

Psychosomatic See *psychophysiologic*.

RA Factor See *rheumatoid factor*.

Reflexology A form of massage usually given to the feet or hands based on a specific theory of innervation and neural reflexes of human organ systems.

Regional Pain Syndrome See *localized fibrositis*.

Repetitive Strain Injury A form of localized fibrositis following repetitive muscle use, usually affecting the neck, shoulder, or arm.

Repetitive Strain Syndrome See *repetitive strain injury*.

Rheumatism From the Greek word *rheumatismos*, meaning "subject to a flux." It is a general term referring to conditions characterized by pain and stiffness of joints or muscles. It includes arthritis and soft-tissue rheumatism.

Rheumatoid Arthritis A form of arthritis characterized by inflammation, pain, and swelling of joints and muscular stiffness.

Rheumatoid Factor A nonspecific factor found in the blood of most people with rheumatoid arthritis. It can be found with other illnesses and in about 5 percent of normal people.

Rheumatologist A physician who specializes in the diagnosis and treatment of the rheumatic disorders.

Shiatsu A form of massage utilizing finger pressure; from the Japanese *shi* (finger) and *atsu* (pressure).

Skeletal Muscles The voluntary striated muscles that are involved primarily in movement of parts of the body.

Soft-Tissue Rheumatism Pertaining to the many rheumatic conditions affecting the soft, as opposed to the hard or bony, tissues of the body.

Somatopsychic Pertaining to the influence of bodily function and illness on the mind.

Strain Injury to a muscle, tendon, or the like by repetitive use or overstretching.

Stress The result produced in the body when it is acted upon by forces that disrupt its equilibrium. The term is commonly used to denote either a cause or an effect.

Subjective Existing or originating within the observer's mind or sense organs, thus incapable of being verified by observation or measurement. As opposed to objective.

Syndrome A collection of symptoms and/or physical findings that characterize a particular abnormal condition or illness.

Systemic Lupus Erythematosus A chronic inflammatory disease that affects the skin, heart, kidneys, joints, and other bodily organs. Its cause is unknown but thought to be closely related to a disorder of the body's immunological system.

Tai Chi Chuan A form of Chinese "soft" exercise emphasizing mental and physical coordination. The motions are dancelike and based on traditional fighting postures and the movements of animals.

Target Organs Bodily organs, such as muscles, heart, and intestines, that respond to stress by abnormal or impaired function.

Temporomandibular Joint Syndrome An illness characterized by pain and tenderness in the temporomandibular joint (jaw joint) and surrounding muscles.

Tender Points Muscle points that are abnormally tender to pressure.

Tendon A cord of dense fibrous tissue uniting a muscle to a bone.

Tendinitis Inflammation of a tendon.

Tension Mental or nervous strain, often accompanied by tautness of the muscles.

Tension Headaches Headaches caused by states of mental, nervous, or muscular strain.

Tissue A collection of similar cells that act together to perform a certain function in the body. The primary tissues are epithelial (skin), connective, skeletal, muscular, and nervous.

TMJ Syndrome See *temporomandibular joint syndrome*.

Tricyclic Antidepressants A drug of a specific chemical structure that has the ability to relieve mental depression. It also has a beneficial effect on the perception of pain.

Trigger Point A muscular point in which applied pressure causes a referral or triggering of pain elsewhere.

Yang The Chinese philosophical concept of the active, masculine force in the universe, as contrasted with the *yin*.

Yeast Any of the various single-celled fungi that reproduce by budding.

Yin The Chinese philosophical concept of the passive, feminine force in the universe, as contrasted with the *yang*.

Yoga A Hindu mystic and ascetic philosophy that embraces specific disciplines, such as those of meditation, exercising, postures, and breathing exercises.

NOTES

Chapter 1

1. M. B. Yunus and A. T. Masi, "Juvenile Primary Fibromyalgia Syndrome," *Arthritis and Rheumatism*, 28(1985):138–45.
2. F. Wolfe et al., "Fibrositis: Symptom Frequency and Criteria for Diagnosis," *Journal of Rheumatology*, 12(1985):1159–63.
3. M. Yunus et al., "Primary Fibromyalgia (Fibrositis): Clinical Study of Fifty Patients with Matched Normal Controls," *Seminars in Arthritis and Rheumatism*, 11(1981):151–71.
4. D. L. Goldenberg, "Fibromyalgia Syndrome: An Emerging but Controversial Condition," *Journal of the American Medical Association*, 257(1987): 2782–87.

5. H. Moldofsky et al., "Musculoskeletal Symptoms
    and Non-REM Sleep Disturbance in Patients with
    'Fibrositis Syndrome' and Healthy Subjects," *Psy-
    chosomatic Medicine*, 37(1975):341–51.

6. T. A. Ahles et al., "Psychological Factors Associ-
    ated With Primary Fibromyalgia Syndrome," *Ar-
    thritis and Rheumatism*, 27(1984):1101–06.

7. J. G. Travel and D. G. Simons, *Myofascial Pain
    and Dysfunction: The Trigger Point Manual*, Bal-
    timore: Williams & Wilkins, 1982.

8. *The Prevention and Management of Occupational
    Overuse Syndrome*, Canberra: Australian Govern-
    ment Publishing Service, 1987.

9. W. E. Stone, "Repetitive Strain Injuries," *The
    Medical Journal of Australia*, 2(1983):616–18.

10. W. R. Gowers, "Lumbago: Its Lessons and Ana-
    logues," *The British Medical Journal* (1904):117–
    21.

Chapter 2

1. Joseph L. Hollander, M.D., ed., *Arthritis and Al-
    lied Conditions*, 7th ed., Philadelphia: Lea & Fe-
    biger, 1966, p. 29.

2. H. H. Stonnington, "Tension Myalgia," *Mayo
    Clinic Proceedings*, 52(1977):75.

3. D. J. Mazanek, "First Year of a Rheumatologist in
    Private Practice," *Arthritis and Rheumatism*, 25
    (1982):718–19.

4. D. Alarcon-Segovia et al., "One Thousand Private
    Rheumatology Patients in Mexico City," *Arthritis
    and Rheumatism*, 26(1983):688.

5. S. M. Campbell et al., "Clinical Characteristics of
    Fibrositis," *Arthritis and Rheumatism*, 26(1983):
    817–24.

6. M. Yunus et al., "Primary Fibromyalgia," *American Family Physician*, 25(1982):115.
7. G. Balint, "A Few Words about Hungarian Medicine and Rheumatology," *Hungarian Rheumatology*, 1987 supplementum, p. 7.
8. D. L. Goldenberg, "Fibromyalgia Syndrome: An Emerging but Controversial Condition," *Journal of the American Medical Association*, 257(1987): 2782–87.
9. D. G. Simons, "Muscle Pain Syndromes—Part I," *American Journal of Physical Medicine*, 54(1975): 289–311.
10. D. G. Simons, "Muscle Pain Syndromes—Part II," *American Journal of Physical Medicine*, 55 (1976):15–42.
11. M. D. Reynolds, "The Development of the Concept of Fibrositis," *The Journal of the History of Medicine and Allied Sciences*, 38(1983):5–35.
12. D. J. Wallace, "Fibromyalgia: Unusual Historical Aspects and New Pathogenic Insights," *The Mount Sinai Journal of Medicine*, 51(1984):124–31.
13. H. Smythe, "Tender Points: Evolution of Concepts of the Fibrositis/Fibromyalgia Syndrome," *American Journal of Medicine*, 81(suppl. 3A) (1986):2–6.
14. P. S. Hench, "The Problem of Rheumatism and Arthritis," *Annals of Internal Medicine* (third rheumatism review), 10(1936):880.
15. J. G. Travel and D. G. Simons, *Myofascial Pain and Dysfunction: The Trigger Point Manual*, Baltimore: Williams and Wilkins, 1982.
16. D. G. Simons, "Fibrositis/Fibromyalgia: A Form of Myofascial Trigger Points?" *American Journal of Medicine*, 81(suppl. 3A)(1986):93–98.
17. H. A. Smythe and H. Moldofsky, "Two Contributions to Understanding of the 'Fibrositis' Syn-

drome," *Bulletin on the Rheumatic Diseases,* 28(1977–1978 series):928–31.

18. E. C. Haugaard, trans., *Hans Christian Andersen: The Complete Fairy Tales and Stories,* New York: Doubleday, 1974.

19. H. Moldofsky et al., "Musculoskeletal Symptoms and Non-REM Sleep Disturbance in Patients with 'Fibrositis Syndrome' and Healthy Subjects," *Psychosomatic Medicine,* 37(1975):341–51.

20. H. Moldofsky and P. Scarisbrick, "Induction of Neurasthenic Musculoskeletal Pain Syndrome by Selective Sleep Stage Deprivation," *Psychosomatic Medicine,* 38(1976):35–44.

21. H. Moldofsky, "Sleep and Musculoskeletal Pain," *American Journal of Medicine,* 81(suppl. 3A) (1986):85–89.

22. D. L. Goldenberg, "Fibromyalgia Syndrome: An Emerging but Controversial Condition," *Journal of the American Medical Association,* 257(1987): 2782–87.

23. F. Wolfe, D. Hawley et al., "Fibrositis: Symptoms Frequency and Criteria for Diagnosis," *Journal of Rheumatology,* 12(1985):1159–63.

24. P. Davidson, *Are You Sure It's Arthritis?: A Guide to Soft Tissue Rheumatism,* New York: Macmillan, 1985.

25. J. F. Jones et al., "Evidence for Active Epstein-Barr Virus Infection in Patients with Persistent, Unexplained Illnesses: Elevated Anti-Early Antigen Antibodies," *Annals of Internal Medicine,* 102(1985):1–7.

26. S. E. Straus et al., "Persisting Illness and Fatigue in Adults with Evidence of Epstein-Barr Virus Infection," *Annals of Internal Medicine,* 102 (1985):7–16.

27. G. P. Holmes et al., "A Cluster of Patients with a Chronic Mononucleosis-like Syndrome: Is Epstein-Barr Virus the Cause?" *Journal of the American Medical Association*, 257(1987):2297–2302.

28. P. M. Boffey, "Fatigue 'Virus' Has Experts More Baffled and Skeptical Than Ever," *The New York Times* (July 28, 1987):13, 15.

29. D. Buchwald et al., "The 'Chronic, Active Epstein-Barr Virus Infection' Syndrome and Primary Fibromyalgia," *Arthritis and Rheumatism*, 30(1987):1132–36.

30. W. G. Crook, *The Yeast Connection*, Jackson, Tennessee: Professional Books, 1984.

31. Comments Requested on Candidiasis Hypersensitivity Syndrome Statement, *The American Academy of Allergy and Immunology: News & Notes* (Summer 1985):12–13.

Chapter 3

1. A. J. Barsky, "The Paradox of Health," *The New England Journal of Medicine*, 318(1988):414–18.

2. M. Friedman and R. H. Rosenman, *Type A Behavior and Your Heart*, New York: Alfred A. Knopf, 1974.

3. D. S. Khalsa, "Stress-Related Illness: Where the Evidence Stands" *Postgraduate Medicine*, 78(6), (1985):217–21.

4. S. E. Locke, "Stress, Adaptation, and Immunity: Studies in Humans," *General Hospital Psychiatry*, 4:1(1982):49–58.

5. R. Melzack and P. D. Wall, "Pain Mechanisms: A New Theory," *Science*, 150(1965):971–78.

6. P. D. Wall, "On the Relation of Injury to Pain: The John J. Bonica Lecture," *Pain*, 6(1979):253–64.

7. Art Rosenbaum, "Playing in Pain," *Sunday Punch,* June 15, 1980.

8. S. Weir Mitchell, *Wear and Tear, or Hints for the Overworked,* Fourth Edition, Lippincott, Philadelphia, 1871; 1887.

9. D. Blumer and M. Heilbronn, "Chronic Pain as a Variant of Depressive Disease," *The Journal of Nervous and Mental Disease,* 170(1982): 381–406.

10. Eric J. Cassel, "The Nature of Suffering and the Goals of Medicine," *New England Journal of Medicine,* 306(1982):639–45.

11. T. C. Payne et al., "Psychosocial Factors Associated with Primary Fibromyalgia Syndrome," *Arthritis and Rheumatism,* 25(1982):213–17.

12. J. I. Hudson et al., "Fibromyalgia and Major Affect Disorder: A Controlled Phenomenology and Family History Study" *American Journal of Psychiatry,* 142(1985):441–46.

13. *Harrison's Principles of Internal Medicine,* Tenth Edition, New York: McGraw-Hill, 1983:70–73.

14. D. S. Scott, "Treatment of the Myofascial Pain-Dysfunction Syndrome: Psychological Aspects," *Journal of the American Dental Association,* 101(1980):611–16.

15. L. G. Mercuri et al., "The Specificity of Response to Experimental Stress in Patients with Myofascial Pain Dysfunction Syndrome," *Journal of Dental Research,* 58(1979):1866–71.

16. *Harrison's Principles of Internal Medicine,* New York: McGraw-Hill, Tenth Edition, 1983:20.

17. *Newsweek—On Health* (Spring 1988):12–21.

18. P. H. Griffin, "Irritable Bowel Syndrome: A Final Common Response to Stress?" *Drug Therapy,* July (1985):45–61.

19. Cecil, *Textbook of Medicine*, W. B. Saunders Company, Philadelphia, Sixteenth Edition (1982): 1932–33.

Chapter 4

1. J. G. Travell and S. H. Rinzler, "The Myofascial Genesis of Pain," *Postgraduate Medicine*, 17 (1952):425–33.

2. A. E. Sola and R. L. Williams, "Myofascial Pain Syndromes," *Neurology*, 6(1956):91–95.

3. J. J. Bonica, "Management of Myofascial Pain Syndromes in General Practice," *Journal of the American Medical Society*, 164(1957):732–38.

4. C. S. Green et al., "Psychological Factors in the Etiology, Progression, and Treatment of MPD Syndrome," *Journal of the American Dental Association,* 105(1982):443–48.

5. R. P. Sheon et al., *Soft Tissue Rheumatic Pain: Recognition, Management, Prevention*, Philadelphia: Lea & Febiger, 1987:2–3.

6. *Acute Conditions: Incidence and Associated Disability,* U.S. 1974–1975, Vital and Health Statistics Series 10, Number 114, USDHEW Publication No. (HRA) 77–1541, 1977.

7. H. Aoyama et al., "Recent Trends in Research on Occupational Cervicobrachial Disorder," *Journal of Human Ergology*, 8(1979):39–45.

8. C. D. Browne et al., "Occupational Repetition Strain Injuries: Guidelines for Diagnosis and Management," *The Medical Journal of Australia*, March 17(1984):329–32.

9. W. E. Stone, "Repetitive Strain Injuries," *The Medical Journal of Australia*, December 10/24 (1983):616–18.

10. D. Ferguson, "The 'New' Industrial Epidemic," *The Medical Journal of Australia,* March 17 (1984):318–19.

11. National Occupational Health and Safety Commission, *Repetition Strain Injury (RSI): A Report and Model Code of Practice,* Australian Government Publishing Service, Canberra, 1986.

12. Hunter J. H. Fry, "Overuse Syndrome of the Upper Limb in Musicians," *The Medical Journal of Australia,* 144(1986):182–85.

13. Editorial, "Overuse Can Stop the Music," *Australian Doctor* (1987):52.

14. Letters to the Editor, *The Medical Journal of Australia,* 144(1986):497–502.

15. G. O. Littlejohn, "Repetitive Strain Syndrome: An Australian Experience," *The Journal of Rheumatology,* 13(1986):1004–06.

16. N. M. Hadler, "Industrial Rheumatology: Clinical Investigations into the Influence of the Pattern of Usage on the Pattern of Regional Musculoskeletal Disease," *Arthritis and Rheumatism,* 20(1977): 1019–25.

17. N. M. Hadler, "Illness in the Workplace: The Challenge of Musculoskeletal Symptoms," *The Journal of Hand Surgery,* 10A(1985):451–56.

18. United States Department of Health and Human Services. "Program of the National Institute for Occupational Safety and Health: Program Plan by Program Areas for FY 1984–1989," NIOSH, Atlanta, Georgia, 1984. DHAS (NIOSH) publication no. 84–107.

19. *VDT News,* P.O. Box 1799, Grand Central Station, New York, N.Y. 10163.

20. E. Schmitt, "New York County Approves Controls

on Video Terminals," *The New York Times* (June 15, 1988):1.

Chapter 5

1. R. M. Bennett, "Studies of Physical Fitness and Exercising Muscle Blood Flow in Fibromyalgia/ Fibrositis." Abstract presented at the American Rheumatism Association Nonarticular Study Group, Houston, Texas, May 25, 1988.
2. S. R. Clark, "Exercise Endurance Capacity in Patients with Fibrositis/Fibromyalgia," Abstract presented at the American Rheumatism Association Nonarticular Study Group, Houston, Texas, May 25, 1988.
3. G. A. McCain, "Role of Physical Fitness Training in the Fibrositis/Fibromyalgia Syndrome," *American Journal of Medicine,* 81(3A)(1987):73–77.
4. *Exercise and Your Heart,* Dallas, Texas: The American Heart Association, 1984:4.
5. J. M. Rippe et al., "Walking for Health and Fitness," *Journal of the American Medical Association,* 259 (May 13, 1988):2720–24.
6. I. Veith, *The Yellow Emperor's Classic of Internal Medicine,* Berkeley: University of California Press, 1966.

INDEX

Index

Index

About the Author

PAUL DAVIDSON, M.D., has been practicing internal medicine and rheumatology in Greenbrae, California, for the past twenty-five years. After earning his medical degree from Albany Medical College, Dr. Davidson was awarded fellowships in internal medicine at the Mayo Clinic and in rheumatology at the University of California. He has served as an associate clinical professor of medicine at the University of California Medical Center from 1974 to the present. Among the prestigious journals he has written articles for are *Pennsylvania Medical Journal; Arthritis and Rheumatism; Journal of the American Medical Association; Prevention;* and *Annals of Internal Medicine.* Dr. Davidson has spoken extensively on both radio and television on the subject of soft-tissue rheumatism. He has recently been appointed the Medical Director of the Chronic Muscle Pain Clinic at Kentfield Rehabilitation Hospital in Kentfield, California.

TAKE CARE OF YOURSELF!

__THE OAT BRAN WAY Josleen Wilson
0-425-11809-6/$3.95
The natural method to lower cholesterol—eat oat
bran. Here's the most up-to-date information
available!

__MIND OVER BACK PAIN John Sarno, M.D.
0-425-08741-7/$4.99
At last! Relief without pain killers and surgery—a
revolutionary alternative that works.

__DR. MANDELL'S LIFETIME ARTHRITIS RELIEF
SYSTEM Marshall Mandell, M.D.
0-425-09355-7/$4.99
The acclaimed patient-tested program that
offers freedom from pain.

__VITAMINS AND YOU Robert J. Benowicz
0-425-12706-0/$5.99
The only guide you need! Tells you what to buy
and when to take it—includes a complete
personal vitamin planner.

Payable in U.S. funds. No cash orders accepted. Postage & handling: $1.75 for one book, 75¢ for each additional. Maximum postage $5.50. Prices, postage and handling charges may change without notice. Visa, Amex, MasterCard call 1-800-788-6262, ext. 1, refer to ad # 239a

Or, check above books and send this order form to:	Bill my: ☐ Visa ☐ MasterCard ☐ Amex	
The Berkley Publishing Group	Card#_____	(expires)
390 Murray Hill Pkwy., Dept. B		($15 minimum)
East Rutherford, NJ 07073	Signature_____	
Please allow 6 weeks for delivery.	Or enclosed is my: ☐ check ☐ money order	
Name_____	Book Total $_____	
Address_____	Postage & Handling $_____	
City_____	Applicable Sales Tax $_____ (NY, NJ, PA, CA, GST Can.)	
State/ZIP_____	Total Amount Due $_____	